TURKEYS

A GUIDE TO MANAGEMENT

DAVID C BLAND

The Crowood Press

First published in 2000 by
The Crowood Press Ltd
Ramsbury, Marlborough
Wiltshire SN8 2HR

British Library Cataloguing-in-Publication Data
A catalogue record for this book is available from the British Library.

ISBN 1 86126 359 7

Acknowledgements
Without the help and patience of my wife Valerie who read and corrected all my original work, plus further help from trusted friends: Janice Houghton-Wallace, Liz Wright, Hans Schippers, Jan Vogel, Chairman of the Dutch Turkey Club, Andy Verelst, President of the Belgium Turkey Club and David Spackman, MRCVS, NDP, this book would have made very slow and painful progress.

Picture Credits
All photographs not taken by the author were supplied by Hans Schippers, Brinsea Incubators, Curfew Incubators, Frances Bassom and T J Smith of Dovecot Studios. My son, Barrie, was responsible for all the preparation and evisceration photographs in Chapter 11.

Line Drawings
Thank you to my daughter, Debra Syme, for her clear and accurate line drawings.

Typeset by Carreg Limited, Ross-on-Wye, Herefordshire

Printed and bound in Great Britain by
J. W. Arrowsmith Limited, Bristol

Contents

Foreword

It is appropriate that this book on the turkey is published in the Millennium year 2000, being exactly 500 years after the first introduction of the bird into Western Europe. Archives in the library of the Royal Palace in Madrid detail Pedro Nino as the discoverer of the bird, on the coast of Cumana, north of Venuzuela in 1499, and on his return to Galicia in 1500 bringing a number which he had purchased for four glass beads each. Cortes found, on entering the capital of Mexico, that 500 turkeys, the cheapest meat in Mexico, were allowed for feeding the vultures and eagles in the Royal aviaries. Long regarded as the monarch of the farmyard, they gradually, over the next 25–30 years, spread throughout the rest of Europe and began to replace the peacock on the dining table, initially in royal palaces and on grand occasions. Originally known as the Indian fowl in most of Europe, the name turkey was adopted in England because they were brought back first by merchantmen who, trading in the Mediterranean, were loosely termed "Turkey Merchants". Archbishop Cranmer prohibited turkey being served in excess of one dish of turkey cocks on State occasions, females being regarded as too precious to cook. That was in 1541, yet only some thirty years later, turkey had become the Christmas dish for farmers, and Thomas Tusser, in his 1573 book, Five Hundred Points of Good Husbandry, remarks of Christmas fare:- "Beef, mutton, and port, shred pies of the best, Pig, veal, goose, and capon, and turkey well drest."

There are three main types of turkey in the world: The Mexican, The North American Wild and the Honduras. It is without doubt the Mexican which first was brought back to Western Europe by the Spanish. It has more black in its plumage than the others. All the early European-bred turkeys were of black colouration, with occasional greys showing up. The Northern Wild turkey of North America is more bronze in colour and by crossing with the Mexican, not least because of its greater size, some wild males weighing 60lb, the original American Bronze, or Point Judith Bronze as they were called, were developed in Rhode Island in the New England area of

USA. From here, the development and spread of the modern large commercial turkey has arisen in a startlingly small period of time.

It is probably time now to recapture some of the fascination and untapped value of other breeds and crosses and this can be achieved on a small scale by following the guidelines offered here. Not only is the turkey regal in its bearing, but the possible colour combinations are breathtaking and worth keeping for that reason alone. It does, however, have the added bonus that even birds which, for one reason or another, fail to live up to the expectations of the breeder in plumage, nevertheless have an additional unsurpassed value as a provider of a succulent, yet healthy, meat. This is attainable at a cost which, relative to other livestock, is within reach of many. It is therefore possible to satisfy the craving of increasing numbers to have an involvement in the "country" without being in agriculture.

With this publication David Bland provides all the technical and practical pointers necessary for interested parties, even those in some urban locations, to take up and develop a useful and fascinating pursuit. Following his earlier, highly successful Practical Poultry Keeping, David has drawn on his years of experience, from breeding through to delivery of final product, to offer the same valuable assistance, in easy to follow language, for those who wish to take up turkey ëfarmingí, be it small numbers of exhibition birds, to larger numbers of more commercially aimed stock. The publication will be of use to enthusiasts around the world and, as he has proved in the past, David also acts as a useful source of information on stock and equipment as well as knowledge, for those keen to start.

I am pleased to be associated with this new guide, which, I feel, has many timely tips not to be found in other manuals.

David Spackman NDP, BVSc, MRCVS

Introduction

The management of small numbers of turkeys has been, and still is, a very important and profitable part of the smallholding. At the present time, it is mistakenly thought that one can only survive by becoming a very large commercial unit, consisting of many thousands of birds. This is completely untrue, as there is a great deal of interest in producing small numbers to supply the Christmas trade and other seasonal markets, and also major functions, and in fact the demand for New York dressed (NYD – meaning 'rough plucked') farm-fresh turkeys in any numbers was a great deal less twenty to thirty years ago. There is less risk from disease when buying birds which are plucked (NYD) than there is from those which have been eviscerated (gutted) and stored in huge numbers for the multiple stores. As regards the latter, because of the system of killing, plucking and evisceration on continuous belts, any infection, if present, will affect large quantities of birds before it is detected. Poultry which are just rough plucked are much safer, as there is far less chance of infection actually getting into the bird.

It should also be borne in mind that these days a very large number of oven-ready birds are imported into Britain. Because there are lower welfare standards in their countries of origin, including less restriction on feed ingredients, they can be produced much more cheaply, making them in turn much cheaper to buy and therefore more profitable to the larger retailer. By contrast, in this country the local farm birds reared in pole barns, straw yards, verandas or on range enjoy a relatively low stocking rate, and so disease and harmful vices are kept to a minimum, ensuring a happier and more contented bird. These turkeys do not suffer from the same stresses as many of their contemporaries, and in many ways are much more attractive to the consumer, even though they will cost more to buy. Furthermore, in the last decade, an increasing number of small producers have diversified into producing turkeys that are kept on organic rations, to supply an expanding niche market. Inevitably this type of bird will be more expensive to buy, but it may provide a better profit margin than the birds on standard additive-free feed. That they taste any better is for the public to decide; certainly it is not proven that they are any more tender or better tasting. Nevertheless, turkeys will be a essential part of Christmas for many years to come, as they are for many other celebrations. And as more and more people become concerned with animal welfare, so the small producer is in an increasingly good position to provide the public with the type of bird they require. To be able to buy such birds from the farm is still a very popular choice with the discriminating section of the public, and many will happily pay that little bit more for a freshly produced, tasty and humanely kept bird.

Some small breeders keep turkeys for exhibition purposes only and it is on these few stalwart enthusiasts that the survival of the many various breeds of turkey will depend. Others just want to keep turkeys as attractive and friendly pets. It is for all these people that this book has been written.

There are various stories bandied about concerning the difficulties of keeping turkeys, but many of these are completely unfounded, and all will be dealt with in the ensuing chapters. It is my fervent hope that anyone who wants to keep turkeys for any reason will be able to. By observing the basic necessities of management they will be able to provide an ideal and stress-free environment – and indeed birds which are kept in optimum conditions are a joy to behold.

Turkeys are indigenous to America and were unknown elsewhere until their discovery on the mainland of the Western Continent in about AD1518. They were shortly to be spread around Europe between the 1820s and 1830s. Different countries lay hold to various claims as to when they appeared on their shores. All were eager to be the first, but turkeys did not initially excite the imagination of breeders, as did the domestic fowl; they were only considered as a bird to eat, and to begin with only cock birds were served, as females were considered to be too valuable for the table. As time went by different colours were produced, via sports and mutations. The first known turkey breed in the UK was the Norfolk Black – or rather, the Black Norfolk as it was known in earlier times. This, and other interesting and vital information, is provided in the ensuing chapters.

— 1 —
Breeds and Breeding

There are many varieties of turkey to choose from, some more readily available than others. If you are breeding for meat, you will need to find out the preferences of your potential customers. If you wish to keep and breed turkeys for exhibition, you will undoubtedly select your favourite strain – though you must be prepared to travel to obtain a few top quality birds from a reputable breeder; however, he or she should be willing to offer help and advice well after the initial purchase. If you wish to keep turkeys just as pets, then choose whatever colour takes your fancy – but again, only buy birds from a reliable source; moreover, take home those that have been selectively bred for health and vigour.

Pied turkey.

BREEDS OF TURKEY

The Norfolk Black

One of the older breeds that arrived from Spain around the fifteenth to sixteenth century; a smaller single-breasted turkey. Compared with the later, heavier turkeys they were not so presentable when New York dressed (rough plucked), or when presented as an eviscerated carcass (oven ready). There are some who consider that these old-fashioned birds are better tasting – and who am I to argue – nevertheless, they are best kept as a small laying and breeding flock to maintain the strain, or as pets or show birds; as the latter they are the second most popular to the Bronze, which is exhibited in greater numbers than any other variety. There are quite a number of turkey exhibitors throughout Britain who breed and maintain many rare breeds which are even more pleasing to the eye.

The Broad-Breasted Bronze

I remember this turkey when it was first introduced on the farm where I was training in the early 1950s. Up until that time we had been using single-breasted black turkeys, and this new bird was a revelation, not only in its transformation from a narrow to a beautiful, wide, broad-breasted turkey, but it also produced a much better meat-to-bone ratio, and thus for the housewife much better value for money. Over the years it has lost ground in commercial popularity to the American Large White double-breasted breeds (*see below*). There is now a small revival of the Bronze, as it is remembered by some as one of the old-fashioned breeds they knew as children, and they consider it to be tastier than the White.

The American Large White

This breed was popular with commercial turkey farmers from its first introduction, primarily because the main problem experienced with black-feathered turkeys, namely that of being left with black stubs on the breast, did not arise; obviously these stubs detracted from the finished article, and were in some cases positively unsightly. I remember visiting a turkey packing station over thirty years ago where a new technique of plucking black-feathered turkeys had been initiated, in which *all* stubs were removed during the process. They thought this method would soon become universal, but instead the American White was introduced, and no other packing

American white hen turkey.

(Left) A champion bronze stag turkey.

American white hen turkeys.

station took up the new process. It worked on the basis of using a higher water temperature in the dip tanks, which actually took off the top surface of skin. This was followed by a very rapid sequence of evisceration, chilling and refrigeration, a process which had to be quick in order to prevent brown blemishes appearing all over the carcass exterior: in normal temperatures, if the surface skin is accidently rubbed off, then the underlying skin almost immediately turns an unsightly brown, making the bird virtually unsaleable.

It was claimed that the Large White produced a faster growth rate, and it could indeed be finished off at about sixteen weeks of age as compared with the Broad-Breasted Bronze which required a further two weeks to mature. This represented quite a saving in feed, though admittedly the White when finished early was lighter; but the market then and now does not usually demand very large turkeys, except possibly over the Christmas period.

The British White

This turkey grows to nearly the size of the Bronze, the stags ranging from between 12.7kg to 17.2kg (28lb to a maximum of 38lb) as compared with Bronze stags which can be anything between 13.6kg and 18.1kg (30lb to 40lb) – just 0.9kg (2lb) difference. In recent years white turkeys have lost a small part of the market, as some more discriminating customers now want to see black stubs on the breast because they believe this indicates that a bird has been reared on a free-range system; in much the same way, brown eggs suddenly grew in popularity, a preference which quickly took over the whole egg market. While this monopoly is not likely to happen with turkeys, serious consideration must be given to the type of bird the customer prefers.

The Bourbon Red

An American breed which originated during the late 1800s and was claimed to be the best table turkey. It enjoyed relative success in America until the early 1940s, by which time it was losing its place in the market as it never in fact lived up to the Large Bronze – there were early claims that it would overtake the Bronze as a table bird. By the 1950s and early 1960s it had become extremely popular in Canada. It is a lovely dark reddy-brown in colour, the primary and secondary wing feathers showing full white when the stag displays; similarly the displaying tail feathers are white from the base, trimmed with reddish-brown around a thick edge. At first sight it may appear to look spotted. Nowadays it enjoys a certain amount of popularity as an exhibition bird, but that is all.

The Buff

This variety was used considerably in producing the Bourbon Red, and when comparing the two there is still a small resemblance in the markings. The hen turkeys are a lovely light buff in colour – although the buff of this variety should not be compared with that of the Norfolk Buff or Buff Orpington poultry breeds. Some would say it is more a cinnamon colour. Although the Buff is chunkier than the Bourbon, it is a smaller bird; stags should weigh out at about 11.3kg (25lb) and the hens at 6.8kg (15lb) – about 3.6kg and 1.36kg (8lb and 3lb) lighter respectively. It is noted as one of the more prolific layers, though its reproductive capabilities are said to be less reliable. It is a very attractive exhibition bird, although the Bourbon pre-empts it as a dual-purpose show and table bird – after all, only the very best are used for the show ring, the owner actually making his living from all the rest as profitable table birds. Perhaps this is why the most popular variety is still the Bronze, whose table qualities for the present-day market are second to none.

There are several other varieties which a prospective owner may wish to keep, depending on the market, for exhibition purposes or because their feather coloration makes them attractive and appealing as pets.

The White Holland

In Holland this was typically a small, compact bird, however its characteristics changed significantly as soon as it was exported to America. Like most of the smaller turkey breeds it matured earlier, and was therefore in considerable demand. It was reputed to have been used to cross with white sports of the larger Bronze to develop the Large White.

The Royal Palm

At exhibitions this turkey attracts a great deal of interest from the public because of its beautiful display of white and black feathering. It is very similar to the European breeds, such as the Pied and the Crollwitzer. The European breeds have been in existence since the 1700s; in America, however, the Royal Palm has only been recognized for the past thirty years or so.

The Nebraskan

This variety has a very distinctive plumage, the under-colour being white with the exception of the back, where the under-colour is grey with black flecks. If this bird is killed before maturity the quills will still be all white. It is a more recent addition and, although double-breasted, was popular as a meat provider only for a very short time. Nowadays it is seen at some turkey shows in the UK.

These are just a few of the many varieties of turkey to be seen, some of which are, or have become, very rare. It is a tricky situation, because if more breeders are to be encouraged to keep even one or two of the many different strains, then more turkey shows must be organized by way of encouragement and incentive. The problem is that for any show to be viable, a reasonable number of entries is necessary – rather a 'catch-22' situation. But as long as those involved in exhibition can be patient, we may yet see more breeders taking on what are extremely attractive species of poultry.

BREEDING

The strength and quality of your breeding stock relies heavily on a strict selection process, and great care must be taken in the selection of both hens and stags. Turkeys have always been bred primarily for meat production, and as such the first consideration must be to look for and maintain the qualities of bone conformation and meat-to-bone ratio. To this should then also be added vigour, vitality and good hatchability.

It is also necessary to maintain the breed standard, and the breeder should be knowledgeable about the varieties he is producing and able to advise purchasers of the essential and desirable characteristics to aim for. This should not, under any circumstances, be taken as far as breeding for show birds only, unless your prime and only intention is to breed solely for exhibition, as this can lead to catastrophic loss of the required commercial potential. This has already happened with the majority of poultry breeds, to such an extent that virtually all the old utility breeds have become extinct, being replaced with aesthetic show breeds which are of no use whatsoever to the commercial section of the industry. The best commercial bloodlines have been lost for ever, as there are very few knowledgeable and experienced poultry breeders left as compared with thirty to forty years ago.

When rearing a batch of growers there must be periods of selection, and both at the beginning of the enterprise and throughout the rearing period, many birds will need to be identified as unsuitable and culled; the 'chosen few' remaining will become the following season's breeders. As soon as possible, fix an identity tag to the wing or a ring to the leg; this identification should be directly related to the record charts of the dam (mother) and the stag, and to the parents' histology.

Early mortality, and those growers which are small or ungainly, should be checked back to the original dam, and if one hen is found to be producing poor quality poults, then any other poults from her should be immediately culled as unsuitable future breeders. Stags should be seen to stand and walk correctly. Those showing signs of any slight limp or staggered gait must be taken out and, as before, cross checked against the dam or dams they were hatched from. These guilty hens should also be immediately excluded from the breeding pen.

Both sexes must be healthy, vigorous and virile, with good body conformation. Nothing short of these standards should be accepted in the following year's breeding pen, even if it leaves you with hardly enough breeders. Such attention to detail will pay off very quickly, so much so that after a few years it will become more difficult to find fault, and then one can confidently sell good quality breeding stock as well as producing first-class young poults. It is unfortunate that within this industry the vast majority of breeders have long gone and, apart for the very few left, it has become uneconomical to carry on breeding as a viable commercial business even on the medium scale. It is now up to smallholders and enthusiasts to support and continue breeding good small flocks of quality birds, and to protect the older breeds from extinction.

Breeding Pens

For a beginner or smallholder, a small pen consisting of one stag to ten hens is the optimum; below this ratio fertility is reduced. Don't expect to have high fertility if you are breeding with pairs or trios – it just doesn't work that way. Provided there is sufficient space to house and run the breeders, they will do better on the ground; however, if space is limited, then a veranda system will need to be considered. Birds with a free-ranging area tend to keep fitter and a little lighter than those housed in more limited conditions. Restricted conditions can adversely affect fertility, more especially towards the end of the season.

The area required per pen of eleven breeders is 4.1sq m (44sq ft) in a house, and 8.2sq m (88sq ft) on the ground. This is providing the outside run is divided into two to allow each side to be rested on a monthly basis. If no such division is made, then allow approximately 1.4sq m (15sq ft) per bird. Each pen will need to be at least 1.2m (4ft) high, and constructed out of 50mm (2in) mesh, and the height will have to be increased to 1.8m (6ft) if the birds are at all flighty and haven't had their wings clipped. When two or more pens are placed side by side they should be screened off by 90cm (3ft) high galvanized sheets or some other appropriate material to prevent the stags from fighting through the fence.

Situate the drinkers and feeders along the perimeter of the fence in such a way that they can be serviced without entering the pen. To do this, use the square stock-fencing along the bottom of the fence where the troughs will be placed so they can be slid through for daily cleaning and refilling.

Breeding pens as well as growing pens can be placed in an old barn with a thick covering of wheat straw as litter. In this environment the use of perches constructed on a very basic frame may be preferred. Nestboxes should not be higher than 30cm (12in) off the ground. A communal nestbox will assist in keeping eggs clean and prevent breakages, in comparison with individual laying boxes, where often more than one bird will attempt to share an individual section. In the communal box they are able to spread out in a peace-

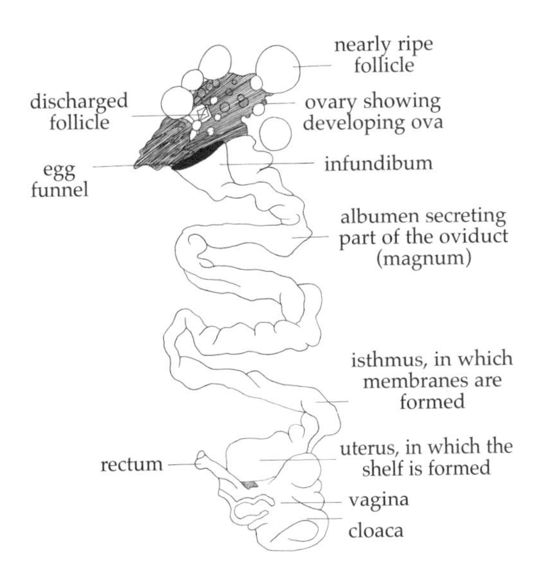

The hen turkey's reproductive system.

ful manner over the larger area, a policy which avoids crowding and, in the summer, possible suffocation. Another advantage is that it is darker, which prevents egg eating and other vices.

Egg Production

Egg production in turkeys is not even comparable with most modern-day poultry, and is only seasonal. Lightweight breeds of turkey are expected to lay between 85 to 100 eggs, medium breeds 50 to 70 eggs, while the heavier breeds will produce about 50 eggs. During the breeding season, the provision of about fourteen hours a day of extra light can be used to stimulate the stags as well as encourage egg numbers. Whether artificial lighting is required depends on what time of the year breeding is expected to take place, and what the normal amount of daylight hours will be at that time.

Mating

Having already selected the best stags, it is always very wise to keep one or two in reserve. Fertility is at its best when using first year stags. As the season comes to an end, working stags can be replaced by the reserve to maintain good fertility and hatchability for the remainder of the season. The small individual breeder may only require a short season, but even so it is wise to keep a reserve stag

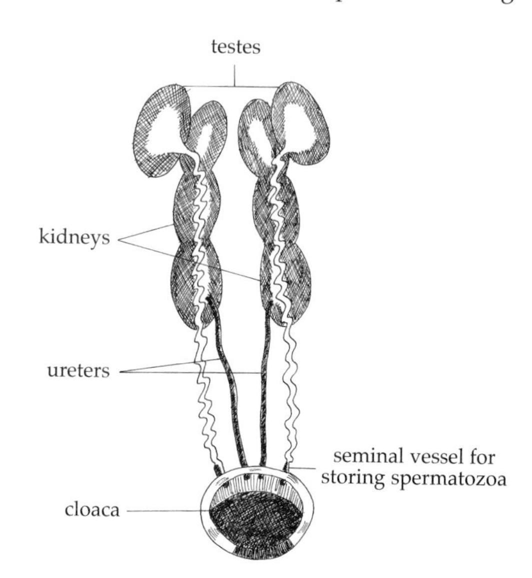

The male turkey's excretory and reproductive system.

in case the other suddenly dies or is killed by a fox or dog. Stags can be used in the second season, but the fertility may be reduced. Hen turkeys will breed well for two to three seasons provided they are kept in a good environment.

Fertilization

After a successful mating, the sperm will head up the oviduct to the funnel area where it is stored. As the ova (yolks) pass through the funnel into the oviduct, they are fertilized. Sperm can be stored for up to several weeks, but by this time it will be very weak. The normal storage time depends on the number of yolks laid; however, on average they can only be expected to fertilize between ten and twelve eggs.

Artificial Insemination

This is another system of obtaining fertility, more especially for the heavier breeds. The stags and hens will need to be kept separately, and semen collected from the stag is administered to the hens. This system is sometimes used as a back-up when poor fertility is occurring in a breeding pen.

Semen is collected two to three times per week, and it needs two people to do this, one to catch and hold the bird down on a padded table or lap, while the other strokes the abdomen, at the same time

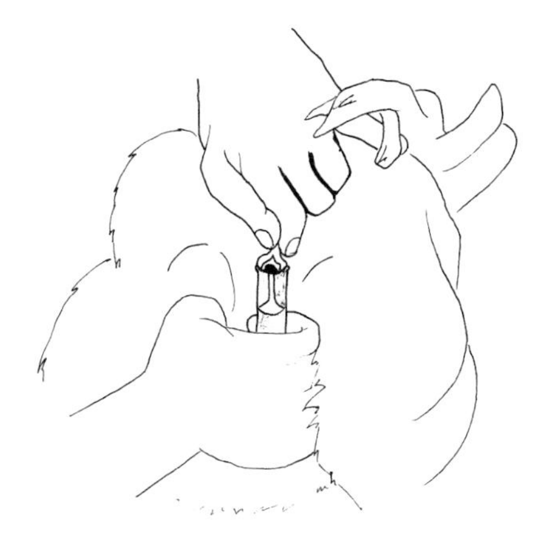

The semen is collected in a small beaker.

The hen is inseminated using a glass tube or a small syringe.

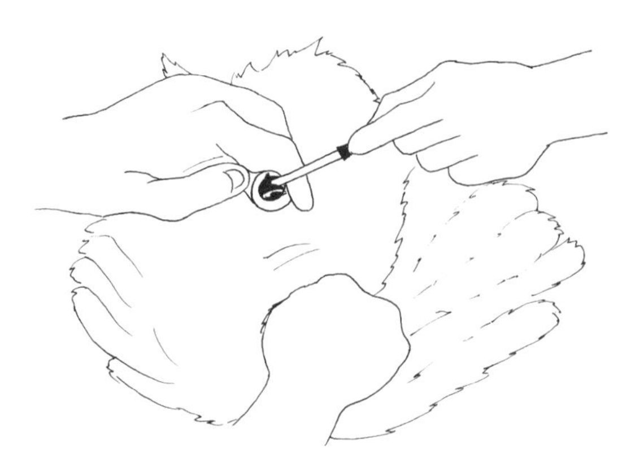

gently pushing the tail up towards the head. As the male responds, the copulatory organ enlarges and partially protrudes through the vent. Once this happens the person stroking the abdomen grips the rear of the organ with the thumb and forefinger, and fully exposes it. The semen is then squeezed out with a short, sliding, downward movement and collected in a small glass beaker or funnel. Stags get used to this treatment quickly, and ejaculate easily when stimulated. A healthy stag should produce 0.2 to 0.5cc (cubic centimetre) per milking.

It is important that the semen is clear from other debris, and to achieve this it has been found that withholding feed some eight to ten hours prior to milking will reduce contamination to a minimum. To avoid stress it is better to carry out this work as soon as possible in the morning so that the birds can eat at the normal time.

The semen produced by turkeys is more concentrated than in most other birds, and can only be kept for a relatively short time: it should be used within thirty minutes. It is therefore essential to be well organized when hens are readily available for inseminating. To obtain the best results, hens should be inseminated twice at four-day intervals as soon as egg production commences. Thereafter every two or three weeks will be sufficient.

To inseminate hens, again two people are required, the first to hold the bird and the second to carry out the work. To inseminate the hen, expose the vent and insert a small needle-less syringe into the oviduct for about 4cm ($1^1/_2$in) of its length; glass tubes or plastic straws may also be used.

Broodies

It is still the natural instinct of birds to go broody, even in this day and age of modern breeding. Broodies are hens that, instead of coming off the nest after laying, will continue to sit there all day and night, coming off only for food and water. Any bird showing such signs should be taken away from the flock, placed in a small cage on slats, and given feed and water for five days. After this period she will have lost the urge to 'brood', and can safely be put back with the others after they have all gone to perch. Hen turkeys are not generally regarded as good mothers, and unless you have one which you know to be reliable, don't attempt to use one – use a broody hen or an incubator instead. (Chapter 5 explains the correct procedures for natural incubation.)

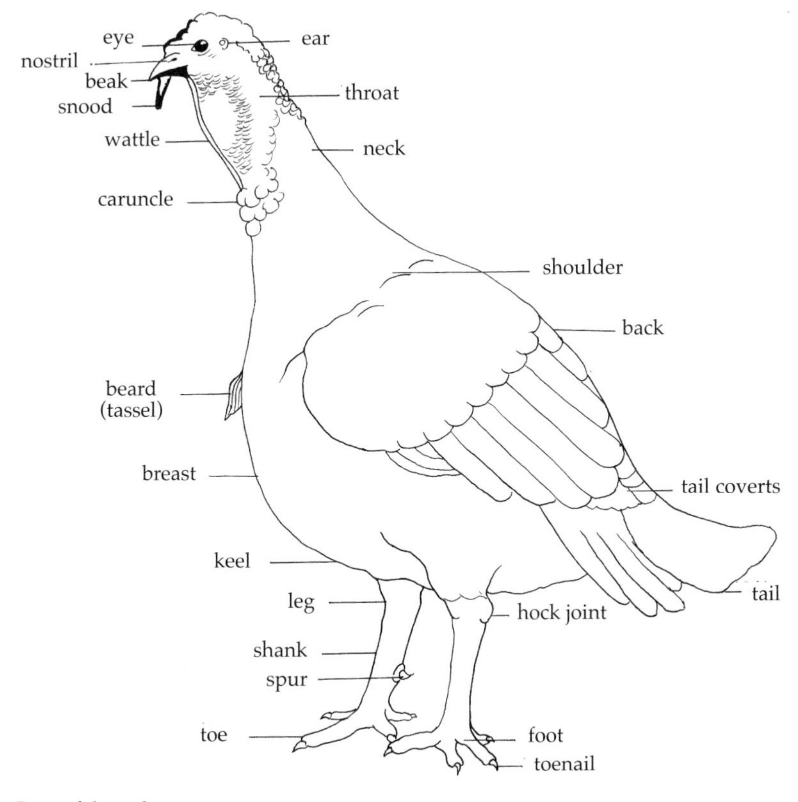

Parts of the turkey.

— 2 —
Brooding

So you have decided on the breed or variety which will best suit your purpose. If you have only a little knowledge, it is better to start with either day-old poults or young growers, generally about four to five weeks of age. If you prefer to see and purchase at the day-old stage, then you must be adequately prepared with proper housing, equipment and feed, before taking delivery or collecting from the farm. The advantage of collecting your young birds is that you will be able to ask those last-minute questions which hopefully will assure you that the preparation you have carried out is correct.

INSULATION AND HEATING

If you already have a poultry house with adequate space, good ventilation and lighting, and that is free from draughts, then the purchase of a further house may not be necessary just for the turkey season. You will know if the house is draughty by the way the young poults group themselves after the surround has been taken down: if they all congregate in one part of the house, you will know that this is the only area free from draughts. They should spread out in small groups over the whole of the house. However, it is self-defeating to try to stop draughts by blocking off any proportion of the ventilation, because the house will then become too stuffy and humid, the litter becoming damp, with possible condensation dripping off the ceiling. This will cause respiratory infections and possible e-coli, leading to early mortality. Furthermore, although young poults like being warm and dry, the windows should be shielded from direct sunlight as this is harmful to them.

During the early stages, turkeys need a good overall room temperature in the region of 21–24°C (70–75°F). The temperature directly under the brooder should start at around 35°C (95°F), and the overhead brooder should be adjusted regularly to the young poults' requirements. If they are too hot they will spread as far away as pos-

Young poults. Ideal temperature – note even spread under heat.

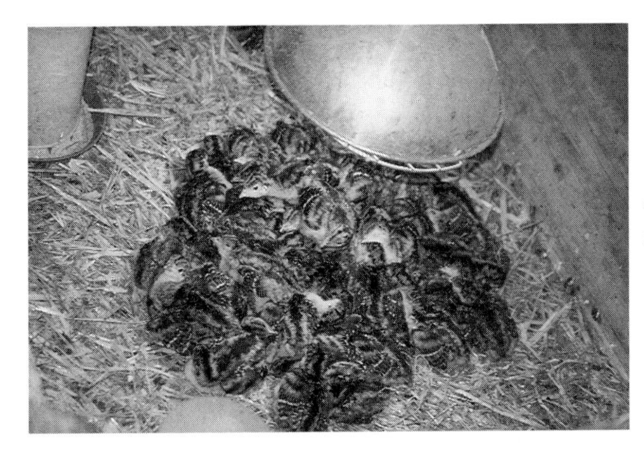

Too cold – brooder needs lowering.

Too hot – brooder needs raising.

sible from the heat source; if too cold, they will group together under the heat element, almost standing on tiptoe. Check them regularly over the first few days and keep adjusting the height of the brooder. An overhead brooder is the most popular and the cheapest for the small producer to purchase. Instead of using the dull infrared emitter, now very popular when rearing other poultry, use a heat bulb which gives off a red light. This will help prevent any very early cannibalism, and it will give out enough light to encourage the poults to feed, besides which it will attract them back under the brooder before they get too cold. There are large, alternative brooding systems for commercial producers, but these are too expensive and completely unnecessary.

The floor should be covered with a layer of clean white wood shavings 7.6cm to 10cm (3in to 4in) deep; this can be added to as the poults become older. The initial circular surround for twenty-five birds need be no larger than 1.2m (4ft) across; any smaller, and they will not have enough room to get away from the heat source. When the poults are about five days old, this surround may be taken down so they can spread over the rest of the rearing area if they want to. Doing so at such an early age will help them develop and feather up quickly and naturally. To contain them in a small area for too long a period will cause them stress, and this will make them less disease resistant, give uneven growth, and possibly cause early cannibalism.

The brooder should be switched on about forty-eight hours prior to receipt of the poults to give time for the litter to warm up thoroughly. In the brooder house or room, place a thermometer about 90cm (3ft) high, and record the max./min. room temperature every day.

THE PROVISION OF WATER AND FOOD

On the first day drinking water must be placed so that the poults cannot help but see it. If poults have been sent to you via a delivery van, especially during warm weather, or if the delivery has been delayed, take them out of the box one at a time and beak dip them before placing them gently under the brooder. To beak dip is to put their beak in water which is just off chill. Hold the poult there for a second or two so that it has felt, and possibly taken, a little water in the beak. When treated this way they are more likely to find the water quickly, so preventing possible dehydration. Be careful that the overall house temperature is not too cold, because then the young turkey poults are less likely to venture out for food and water with the result that four to five days later a few will appear to have shrunk, with their small wing feathers sticking out. These chicks will die. It is more important for them to drink than eat during the very early stages, so always ensure that they can see clearly the drinking points provided.

Some time ago, in the early fifties, turkeys were considered to be difficult to rear, largely because day-old poults would often refuse to eat. To get them going, chicks a few days old would be put with them to show them how to feed; later on, at least one large feed company included red chips with their normal turkey crumb diet. Turkeys, like chickens, are attracted to red more than any other colour. Feeders were also sometimes painted red. However, the real reason for these problems was that, in those days, chicks were started off in small outdoor brooders whose only light was a dim oil heater in the centre, and as natural perching birds this system rendered them semi-blind – it would have been rather like trying to feed roosting birds at dusk. Once the production of turkeys became more commercialized, most poults were started in broiler-type

Young poults with drinker near heat source.

houses which provided good, evenly distributed lighting; like this, poults could see properly and the problem disappeared. Today turkey rearing is now much better understood so that both small and large producers are able to rear extremely successfully.

For the first few days use shallow feeders, such as new keyes trays (egg trays) or even chick-box lids; there is bound to be more spillage, but poults will pick up more feed by this method and this will reduce early mortality – the loss of feed will be small compared to the number of chicks you might otherwise lose, making the method viable. Once poults have settled down, change over to waste-free feeders.

EARLY MANAGEMENT

Turkeys are very susceptible to cold in the early stages, and during the colder months should be kept in until they are about 8 weeks old. If they are reared during the summer months they can be allowed out during the day time at about 5 weeks of age, although they must have adequate protection from the sun. At all times the litter in the house should be kept dry and friable. At 12 to 14 weeks, after their period on range, they may be transferred to more permanent quarters until they are ready for preparation. By this time the few to be kept as pets should be well imprinted with their owners – from 8 to 12 weeks they will probably be following their 'people' around like a pet dog.

After the first five days of age, birds will need enough space if they are to thrive, and pens should be organized according to the following parameters: up to 4 weeks old, 0.0929m^2 (1sq ft) per bird; 4 to 8 weeks, 0.1394m^2 ($1^1/_2$sq ft) per bird; 8 to 10 weeks, 0.1858m^2 (2sq ft) per bird. It is better to rear in groups no larger than 100, although batches of fifty or less seem to rear better.

Broody turkeys usually make very poor mothers as the modern domesticated varieties can be exceedingly clumsy; broody hens, on the other hand, are ideal, although sadly it is now rarely possible to get hold of such a bird. There is, and always has been, a temptation to rear very small batches in cardboard boxes with an overhead heater. This can be very unsatisfactory unless there is sufficient area for chicks to get away from the heat source, and it is essential to increase the floor area after they are five days old. Keep a careful eye on them to ensure they do not stray away from the brooder heat and have difficulty in finding it again. As a rule, the heater temperature can be reduced by about 6°C (10°F) each week, but it must be appreciated that this is only a general guide. Each batch of chicks can vary, and the poults themselves are the best guide. Basically it depends how warm the ambient house temperature is, because this will dictate how high you can raise the brooder. Some books recommend that the brooder is raised only on a daily basis, but this does not necessarily constitute the best management procedure. Basically, when making any adjustment, always observe the reaction of the chicks, and readjust the brooder if necessary; during the first weeks, several adjustments to height may be made during the day. The most important adjustment is the one made just after the chicks have settled down for the night: then the brooder should be positioned so that there is a clear space about the size of the average saucer in the centre; this will allow for any further drop in room temperature during the night.

Before you remove the brooder surround, move the feeders and drinkers just a little away from the centre of the brooder until they are at the edge of the surround; do not move them at the same time as enlarging the floor area, because the poults will not find them, even though they may be only 30 to 45cm (12 to 18in) further away – turkeys are blessed with an extremely small brain, and should be treated accordingly. So move their feeders only a little the day after the surround has been removed, and then just a little bit each day until they have reached where you want them to be. This is assuming that the poults will be staying in the same house until fattening has been completed. If they are to be moved to more permanent quarters, then provide a similar layout during the early rearing stage to help them settle down quickly, so preventing unnecessary stress. If they automatically know where to find their feeders and drinkers, they will accept their new quarters with few, if any problems.

During the summer months the brooder can be turned off when the poults are between 5 to 6 weeks of age. During the colder part of the year, more especially if the house temperature is also low, it may be necessary to maintain the brooder heat until they are about 8 weeks old. In both cases do not remove the brooder immediately after turning off the heat as some will find it a comfort to sit under, even without heat. In other words, try not to be too clinical with your management changes; work with the birds and make any necessary alterations gradually, always being prepared to reverse any decision immediately if the poults appear stressed. That little extra heat can make a considerable difference to successful rearing and the quality of the birds. Having said that, there is no advantage in keeping the brooder heat on too long.

Using a Broody Hen

There are some who may prefer a broody hen for both incubation and hatching. If a broody is available from your own, or a friend's, laying flock, she must first be settled in a small brooder coop or similar box. It only need measure 30 × 30 × 30cm (12 × 12 × 12in), but it must have good strong sides, roof and back, and if not a solid floor then one constructed of 13mm ($^1/_2$in) mesh. Cover the floor with about 7cm (3in) of soft, clean, dry straw or soft white wood shavings. Settle the broody hen on dummy eggs in her new quarters two or three days beforehand, letting her out only once a day to feed and drink. She should already have been thoroughly dusted with a good louse powder such as 'Eradicate'.

When the day-old poults arrive, place them gently under her and keep her shut in the coop in semi-darkness for two or three hours, to allow time for her and the new poults to settle down. Once you are satisfied they have become accustomed to their new environment, you should hear the broody talking to them. Place their feed and water immediately in front of the coop, then open up the front; at this stage all should be protected by a small run, in the first instance to prevent the poults from straying too far on their own, if the coop is outside, and also to stop predation by crows and magpies. For at least the first two days it is advisable to restrict the broody to the coop – use vertical bars that are close enough to keep her in, but just wide enough for the poults to pass through; she will still be able to teach and encourage them to drink and eat, but without upsetting the feed and water containers.

In general, chicken and turkeys should not be kept together because of the risk of the disease known as 'blackhead': although it is rare for chickens to become ill and die, all the evidence points to the fact that chickens somehow act as carriers, and even small groups of turkeys are more likely to go down with blackhead and die than when they are kept completely separate from other poultry. Having said that, they seem to rear very well under a broody hen who is far less clumsy than a hen turkey.

Poults reared in this way can be separated from their foster parent at 5 to 8 weeks of age, again depending on the weather. If, however, they are still with the broody at over 5 weeks of age, you will see that at night-time the poor hen is often left suspended over them as they are now larger than her, and so her feet cannot comfortably touch the ground!

The average size broody hen is capable of rearing up to twelve poults, and a hen turkey twenty.

3

Rearing and Systems of Housing

Once the birds are successfully off heat, unless they are to stay in the shed they have been brooded in, they will need to be taken to further accommodation. Basically there are four systems available, and the individual rearer will choose whichever suits him or her best: these are free-range houses, fold units, verandas and pole barns. However, I would not recommend that poults much younger than 8 weeks of age are placed in any of these.

VERANDAS

In the fifties and early sixties verandas represented the most popular system for rearing turkeys in small groups; like this, they were considered easier to manage. The main problem was that, as commercial flocks became larger, they were too labour intensive – although this is not really applicable to the small rearer. The main advantage of the veranda is that it can be built at home by anyone who can use a hammer and a saw, birds of different ages can be kept separately and stags can be reared independently from hens. Furthermore, birds reared and kept in the open tend to be healthier and more disease resistant than those kept intensively. Finally, these units take up very little space – one could fit very well into a back garden of reasonable size. The only forseeable problem might be the noise that turkeys make when they are disturbed, and it might be worth taking the trouble to discuss the situation with close neighbours; you could suggest that at Christmas time they might like to purchase a guaranteed home-raised turkey, in the full knowledge that it has been kept humanely and not been fed on anything of dubious origin, as are so many of the imported birds bought cheap from abroad by less discriminating supermarkets. It is possible you will receive full support for your little enterprise.

A turkey veranda.

A veranda measuring 1.5 × 5.5m (5 × 18ft) can house up to thirty small turkeys, or fifteen of the heavier strains. It stands 60 to 90cm (2 to 3ft) off the ground, and there is no need to clear away any manure until after the unit has been evacuated. For the roof, cover it with chicken wire, then lay cheap lining felt over the top of this, and add a further layer of netting over the felt to keep it in position and prevent it from being ripped off by the wind. All the birds can be seen and inspected easily, and any sick or sorry poult taken out without upsetting the others too much.

When the young birds are first introduced to their new accommodation, place a layer of straw over the slats. This will drop through gradually, enabling the poults' adjustment to a slatted floor is also a gradual process, thereby eliminating any possible stress that a sudden change from solid floor to slats might cause. Under the unit lay out straw at a depth of 30cm (12in): this will help knit together the droppings that accumulate in the first few weeks. Add more straw as and when required. The roof of the veranda will help to keep the droppings dry, and in doing so will prevent them from smelling too much; it is mainly wet droppings that give off a strong smell. Once the birds have been moved, the litter in its dry state can be moved and stacked, if it is not wanted immediately. Keep it covered at all times and it will rot down quickly and be much easier to handle. Beware how you use it, however, as it will have a very high nitrogen content. It can be used profitably by digging it in during the winter months, or by trenching and covering with approximately 8cm (3in) or so of soil.

FOLD UNITS

These are of similar construction to the chicken fold, but higher. The house has a slatted floor, so that when it is moved to a fresh patch of grass each day, the droppings are left behind, evenly distributed on the ground. As with verandas, commercial labour costs these days would make this system too dear; for the small producer, however, with plenty of good grass on a level pasture, turkeys rear very well – they also tend to eat less of the standard ration, as good grass will account for some of their appetite. Furthermore, it is a good free system of fertilizing the ground at no extra expense, and sheep and cattle do well when grazed on grass fertilized in this way. However, it is of crucial importance that no chicken have been run over this ground for the previous six to seven years because of the risk of blackhead; and if they have, then the grower's feed will have to be medicated to offset the risk.

Allow at least 0.4sq m (4sq ft) per bird up to killing age. A fold unit measuring 1.2m wide by 1.2m high by 3.6m long (4 × 4 × 12ft) is suitable for twelve turkeys.

It is important that poults in this – as in every other – system are provided with a quality, well-balanced feed on an ad-lib basis. In

Bronze turkey adults in a pole barn.

preference, it should be in the form of a coarse ground mash, though unfortunately this is difficult to obtain these days. Feeding pellets often leads to boredom and its associated vices, and to minimize this tendency hang up cabbage stalks or other greenstuff; and if these are not readily available, then suspend a couple of lengths of chain from the roof.

Young poults coming to the fold-unit system from a solid-floor environment might find it difficult to adjust to a slatted floor; in this case a sheet of brown paper laid over the slats for the first day or so – or for as long as the paper lasts – will help them to do so more easily. The slats should be 4cm ($1^1/_2$in) wide, and 3cm ($1^1/_4$in) thick. Do not put them on chicken slats which are too thin for heavy turkeys and will cause feet blisters and subsequent lameness; such birds lose weight and will be picked on by others. There should be no danger of slats producing breast blisters, as might happen when these heavy birds are encouraged to perch.

POLE BARNS

Pole barns are sometimes referred to as turkey yards, though I would suspect the only difference is that some turkey yards are constructed as an open, yet fenced run, with the poults having a large covered shed. Pole barns became popular when second-hand telegraph poles were cheap and plentiful, and commercial units were then relatively small. These could be constructed with either a pent-style roof (sloping) or an apex, the latter at the time being the most popular.

Rearing turkeys in a pole barn.

White and bronze turkeys being reared in an intensive house.

Poles are dug in the ground to support the roof. Netting of 5 × 5cm (2 × 2in) mesh is stretched along the side and ends, and fixed to each upright. Corrugated sheeting is also fitted along the bottom of each side and at both ends, giving wind and weather protection. The floor of a pole barn is the soil it has been erected over, and is covered with at least 15 to 23cm (6 to 9in) of clean, dry wheat straw. Never, ever use damp or mouldy straw, as this may be responsible for a costly outbreak of aspergillosis. If the barn has been constructed to house 100 or more birds, it is a good idea to lay rows of straw bales from virtually one side to the other – that is, leaving a passageway on each side; the lines of bales should be about 3m (10ft) apart. These will keep poults better separated, and in the event of bullying, the bird that is being picked on is more able to get out of the way by running round the row of bales, so that the bully soon loses interest. When birds are ready to be marketed, a couple of simply constructed catching frames should be used, to segregate small groups for loading.

Poults are usually put into this system at about 8 weeks of age, and for the first few days should be kept down at one end until they have adjusted to what will seem to them at first to be a completely alien environment. Once they have settled down, take away the restraining frames to allow them the whole of the floor area.

If the weather is still cold when poults are due to be put out into the pole barn, hang a lightweight frame covered with hessian or paper bags over half the restricted area; suspend this frame a good metre (4ft) high off the litter so they can settle under it at night. This false ceiling helps them to retain a sufficient amount of their body heat to prevent them from getting too cold and, as a consequence,

American white and bronze turkeys being reared together.

crowding up to each other. As with all other systems, when birds are moved, it is vital that their keeper watches them carefully and changes any part of his/her standard management procedures to suit the different requirements of each batch. It is a mistake to make management procedures too rigid just because they suited the birds the previous year. Weather conditions will always vary from one period to another, just as each batch of poults will react differently to previous batches.

FREE RANGE

If you are planning to rear turkeys on a free-range system, consideration must be given as to whether chickens have recently been kept on the land; if they have, turkey poults will be put at risk of developing blackhead, and their feed will need to be medicated. It doesn't matter how small the flock is, the risk is just the same. The pasture must have a good growth of short grass, be well drained, and have sufficient shelter against sun and wind. Windbreaks and overhead shelter can be man-made, but it is very important that they are provided.

Turkeys are like geese whose instinct it is to flock together, and just like geese they are easily rounded up and moved to wherever they are required to be, whether this is to a fresh area of pasture, or back to a solid and fox-proof house in the evening. It is a mistake to think that once turkeys are nearly fully grown the fox will not be able to attack them: he will, and he will undoubtedly wreak havoc, killing many just for fun while panicking the rest.

Temporary pens can be put up quickly and easily using electric poultry netting. By grazing one section at a time, the ground can be kept in a good state and the birds will be more contented and healthier because of it. Turkeys are naturally stupid, so their manager must keep an eye on them during the day, especially when first putting them out, to check whether they are making proper use of the shade and windbreaks. As with all the other systems, sick birds must be separated as soon as they are spotted, and kept in some form of individual confinement for treatment, or to recover from injury, be this cannibalistic or otherwise.

By natural instinct turkeys would not range very far away from their food or water, so these containers must be moved with them. The house provided for them to be kept in safety at night can have an earthen floor covered with straw, and should be large enough to keep them in during days when the weather is very inclement, such

Strawyard system.

Bronze turkeys in a covered run.

Feather-pecked breeding hen.

Applying anti-peck spray.

as heavy rain. Some provide such houses with perches, but these are not recommended for heavy birds. A mobile house or houses have the advantage that they can be moved with the birds to each new patch of grazing.

FEEDERS, DRINKERS AND GROWTH RATES

Turkeys grow extremely quickly over a relatively short period of time, and so it is important to understand their daily requirements. As they grow, so they can be more wasteful in their feeding and drinking habits. Feeders should be relatively waste free, and drinkers should be increased in size. Over the last two or three years we have seen a comeback of the 13.6ltr (3gal) bucket drinker – a happy event, as this is easy to fill and carry without spilling. Water is very important for weight increase, and there should be sufficient drinkers so you are not having to run back and forth constantly filling them, as this may not always be convenient. During my second year's poultry training, my manager taught me always to prepare ahead so that should some catastrophe occur at any time it could be dealt with without any birds suffering for lack of water or food. This meant that buckets or bags of feed were always prepared in advance, and there was always an extra drinker or feeder on site should the operator be delayed for some reason. In other words, always think and prepare a day ahead.

For ease of calculation, the table below has been worked out on 100 birds.

Daily Water Consumption Table of Heavy Turkeys
Litres and Gallons per 100 birds

Age in weeks	*Water consumption*	*Age in Weeks*	*Water consumption*
1	4.55ltr (1gal)	11	63.64ltr (14gal)
2	9.09ltr (2gal)	12	68.18ltr (15gal)
3	13.64ltr (3gal)	13	72.73ltr (16gal)
4	18.18ltr (4gal)	14	75.01ltr (16.5gal)
5	22.73ltr (5gal)	15	77.28ltr (17gal)
6	27.28ltr (6gal)	16	75.01ltr (16.5gal)
7	34.10ltr (7.5gal)	17	75.01ltr (16.5gal)
8	43.19ltr (9.5gal)	18	75.01ltr (16.5gal)
9	50.00ltr (11gal)	19	75.01ltr (16.5gal)
10	56.83ltr (12.5gal)	20	75.01ltr (16.5gal)

Please note the above table is only a guide to management and, as has been stated previously, allow and provide for more drinkers and feeders than any standard guide. Such a policy not only protects stock from possible shortages due to matters sometimes beyond your control, but will also help reduce the incidence of possible bullying. Always place them over the whole area, too, as this helps to keep poults well apart.

Tables for bodyweights and feed consumption are shown in Chapter 8.

I have seen turkeys reared and kept until fattened in crude houses constructed of straw bales. These are made proof against foxes by stretching wire netting around the whole outside circumference, and roofing with corrugated iron. They are usually constructed for very small numbers of birds, and are not only very cheap to build but can also be burnt at the end of the season, thereby preventing any possible disease build-up for the following year. Their main disadvantage is that rats and mice find them ideal living quarters, with feed readily available and a warm cosy nesting area between the bales. Even if the owner can put up with this type of vermin, rats and mice are dangerous disease carriers and should not be tolerated living side by side with any type of poultry, whether for meat or egg production.

4
Equipment

EQUIPMENT FOR THE BROODER HOUSE

We have already established that a brooder house needs to be draught-proof, yet ventilated and warm. Now we must consider the type of equipment we need for the enterprise to succeed.

Heating

Firstly an overhead heater is required. Gas brooders are not really economical for the small rearer, so an overhead electric brooder becomes the ideal piece of equipment; these are purchased with a shade that deflects the heat downwards. There are basically two choices of screw-in heating element: a porcelain infra-red dull emitter which gives off no light at all; and an infra-red glass heater which provides a dull red light. If the former is used an alternative light source is required, normally a window, but at night until the poults are off heat a 15-watt pigmy bulb suspended by the side of the brooder will attract them to the heat source. If a red glass light is cho-

An infra-red brooder, complete (left) and in sections.

sen, then extra light will be needed for fourteen hours a day in the form of a 60-watt bulb – unless the windows in the brooder shed provide sufficient natural daylight that they can see to drink and eat.

Infra-red heaters are preferred during the early stages of rearing because young poults are less likely to become lost or chilled, and the owner can see them more easily to check for any untoward problems that may occur and also to remove the odd dead poult. The only problem with using a glass unit is that, when in use, if cold water is accidentally flicked on to the heater the glass will disintegrate. But with proper adjustment and care this is most unlikely to happen with very young turkeys. As they grow, so a change to the dull emitter infra-red may be required (*see* previous chapter).

Two standard-size sheets of 2.4 × 1.2m (8 × 4ft) of hardboard sawn lengthways will make a useful and necessary surround when pegged together.

Feed and Water Containers

Feeders for the first day or so need only be new keyes trays – the trays that eggs are placed on – after which small troughs or tubular feeders resting on the floor are preferred. As the poults get older, in order to avoid wastage the tubular feeders should be raised off the ground, so that the base of the feeder is level with the top of the bird's back. Also, keep the centre cone as low to the feed base as possible to prevent the feeding pan filling up too much; otherwise the poults will hook out the feed, leaving half of it wasted on the floor.

Drinkers should not be too large at first, but there should be plenty of them; small, relatively cheap plastic founts which serve this purpose are easily available. As the poults start to grow, then a fount on a small stand of bricks will be necessary.

EQUIPMENT FOR GROWERS

The next stage of equipment will be needed when the poults are about 6 weeks of age.

Water Founts and Feeders

By this stage larger and probably fewer water founts are required. In the past few years, we have seen the return of the bucket drinker, which is so very handy and practical to use for turkeys kept in free-range conditions. This can be filled easily from a tap, and it is then car-

Alternative young poult feeders during brooding.

Choice of drinkers from day-old through first four weeks.

Adult drinkers: (left to right) automatic field drinker; 8ltr hydroplastic drinker; 13ltr hydroplastic drinker; galvanized bucket drinker.

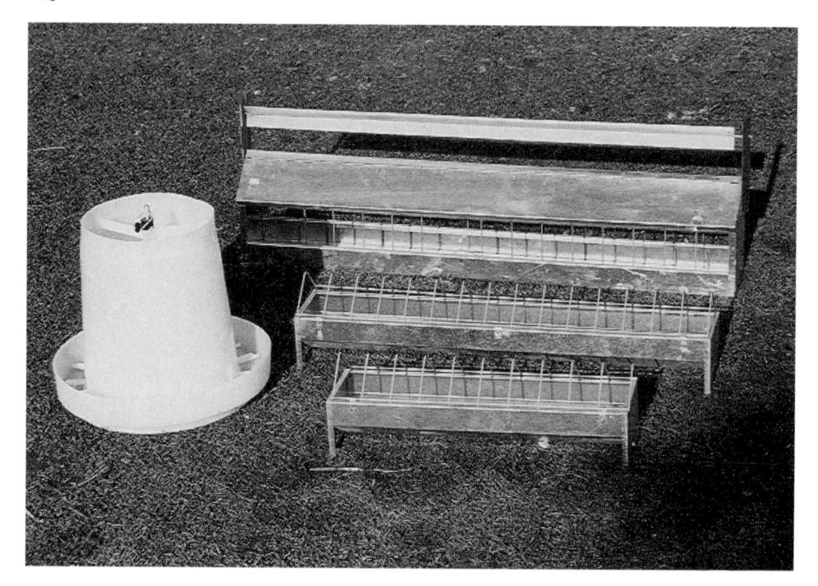

Adult indoor troughs, outdoor trough and hanging tubular feeder.

ried as a normal bucket to its destination where it is laid on its side – and without any loss of water! Larger founts can be difficult to carry any great distance, as sometimes the handles cut into the hand. However, it is not too difficult to make one's own water troughs to hang on verandas: these consist of two blocks of wood as end sections with a tin shape like a very flat U tacked to each end with large felt nails. Squeeze a little Bostik or similar sealer on the three sides of each wooden end; once filled, the wood enlarges and a perfect seal is made.

This is also the time to change to greater capacity feeders. These can be home-made of wood, or they can be manufactured large tubular feeders. Feed represents some 75 per cent of the cost of a growing and breeding turkeys, so whatever type of feeder you choose, it must be waste free.

Floors

Slatted floors are constructed of slats 3.8 × 4cm ($1^1/_2 \times 1^1/_4$in) deep, nailed 2cm ($^3/_4$in) apart on a 5 × 3.8cm ($2 \times 1^1/_2$in) frame; this covers a small area for a small number of poults. If slats are to be made for the large veranda then it would be preferable to use a 3.8 × 7.6cm ($1^1/_2 \times 3$in) framework.

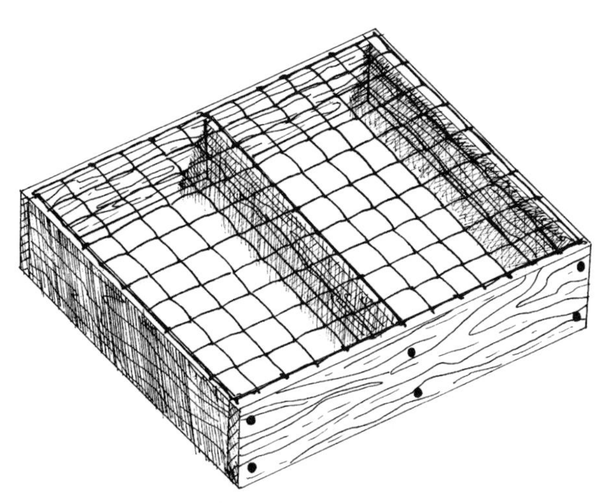

Wire water stand.

EQUIPMENT FOR BREEDERS

Perches

Perches may be preferred for medium- and light-breed turkeys, and they should be constructed on the level and not slanted as seen in journalistic-type books. The reason for keeping them level is that turkeys, like chicken, love to get to the highest perch, and so those constructed at different levels will almost certainly lead to unnecessary squabbling and stress. Some birds will also begin to roost earlier than the others to get on the 'top perch'. Perches which are level prevent this unnecessary competition, and birds do live more harmoniously together.

Perches should be made of 5 × 5cm (2 × 2in) sawn timber. Prepared wood is not only slimmer and therefore weaker, but it is more difficult for the poults to get a grip on. If they slip as they perch this can give rise to simple lameness and at worst a broken leg or wing. The perches should be mounted 76cm (2ft 6in) apart on a 10 × 5cm (4 × 2in) frame at 1.2m (4ft) intervals.

Nestboxes

Nestboxes can be either communal or with individual compartments, depending on the preference of the breeder. The measurements for individual nestboxes are shown in the diagram (*overleaf*),

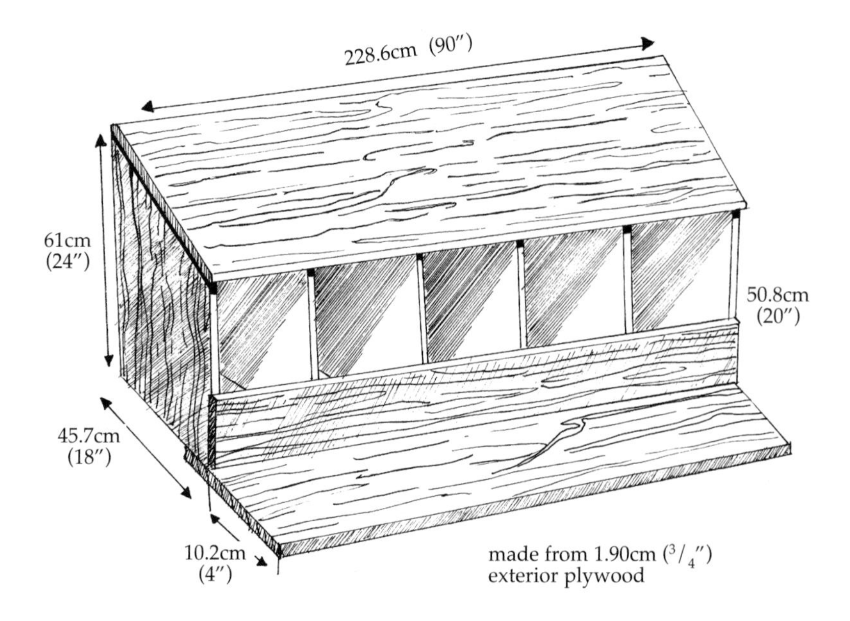

A standard individual nest box.

the only difference in constructing a communal one being that there are no internal individual separations, but instead the front is fixed, and there is a 30 × 30cm (12 × 12in) opening at one end where the hens can pass in and out. The lid, which is the main part of the roof section, needs to be hinged 15cm (6in) from the back, the back part giving the whole construction rigidity. Eggs are collected by opening the lid, and birds can be gently lifted out without disturbing the others. With such a large top-opening access the nestbox can be easily cleaned out, and the shavings replaced each week. Nestboxes are normally constructed using 19mm ($^3/_4$in) exterior ply, a cheap and easy form of material.

When visiting fellow breeders observe what equipment they are using, how it is being used, and whether there is any obvious wastage. It is always cheaper to learn from others – as long as mistakes are also taken into consideration. The reader will find that there are some breeders with very fixed ideas; but then they miss the fun of the continual learning process – which, with livestock is never ending.

5

Incubation

The reason for keeping turkeys is not just to eat them, but to enjoy seeing the various breeds, being in their company, and of course for breeding replacements. Over-fat turkeys will be difficult, if not impossible to breed from; a feeding programme is dealt with in Chapter 8. Turkey eggs can be successfully hatched under hens and in incubators, but not often under turkeys themselves. The care required and the satisfaction of a job well done is a very exciting part of breeding turkeys. For the purpose of natural incubation keep a few pure-breed hens such as Rhodes, Wyandottes or Light Sussex, or better still, cross large fowl hens with a Silkie cock. The resultant progeny are excellent mothers, will become broody at the drop of a hat, and can cover a large number of eggs.

Curfew turkeybator I. 100-egg capacity.

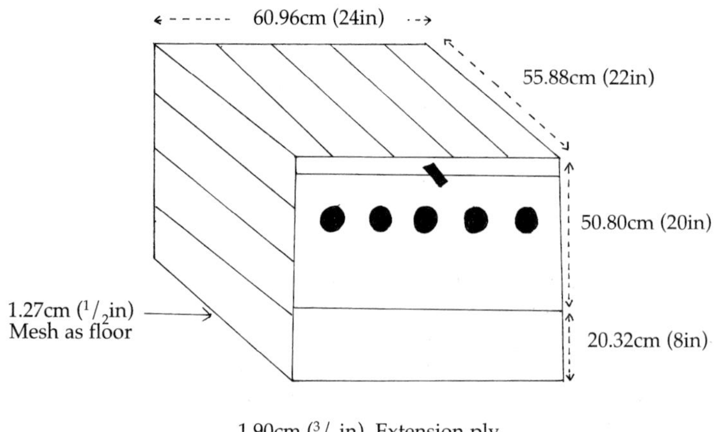

Natural incubation nestbox.

NATURAL INCUBATION

Around the time you intend to breed replacements, start saving your fertile eggs. Do not save every one you collect, but those which are of a good standard size, and well shaped with a good quality shell. They should also be clean. Write down the collection date on each egg, and place them point down in a keyes tray (egg tray). Reinforce the base with an extra tray to prevent the eggs from cracking against one another when the tray is lifted, and place in a cool clean room, away from windows and doors. The ideal temperature for storage is 10–12.8°C (50–55°F). Do not keep any eggs more than fourteen days as hatchability will be adversely affected, and carry out a further selection at the time of setting; thus you will be using only the very best quality eggs when a satisfactory broody has been found. If you reach the number of eggs you intend setting before you have found a broody, then each day replace the oldest eggs with the fresh ones you have just picked up; use the discarded ones for household purposes. It can now be better appreciated why more people prefer to use an incubator: simply because they can plan times of saving and incubation.

Once a turkey or hen has become broody, leave her for four to five days to settle before attempting to set any eggs under her. It is better to put her in a small, specially constructed coop set aside for this purpose where she can settle and brood her eggs. A turkey should be able to incubate between fifteen to eighteen eggs at a time, though this is relative to bird and egg size.

A broody coop should consist of a box of approximately 0.186sq m (2sq ft), the base of which is covered with 13mm ($^1/_2$in) galvanized mesh. Alternatively, sit the coop on an appropriate mesh panel. To form the perfect floor covering, cut a turf the same size as the base of the coop. Turn it over, grass facing down, and scoop out a smooth, concave area. Once this has been done, turn it grass uppermost and place it in the bottom of the coop; the eggs will then always roll to the middle, and as the hen shuffles them about none will be left at the side of the nest and become chilled. This is why the nests of most wild birds are shaped like this. Place a layer of straw on top of the turf, and then put two or three artificial eggs in the nest.

The coop should have a hinged front, and this is only lifted up once a day when the broody comes off to feed and drink. Ensure that there is adequate ventilation in the box; several 25mm (1in) holes drilled around the top should suffice. Keep the coop in a quiet

First-quality turkey egg.

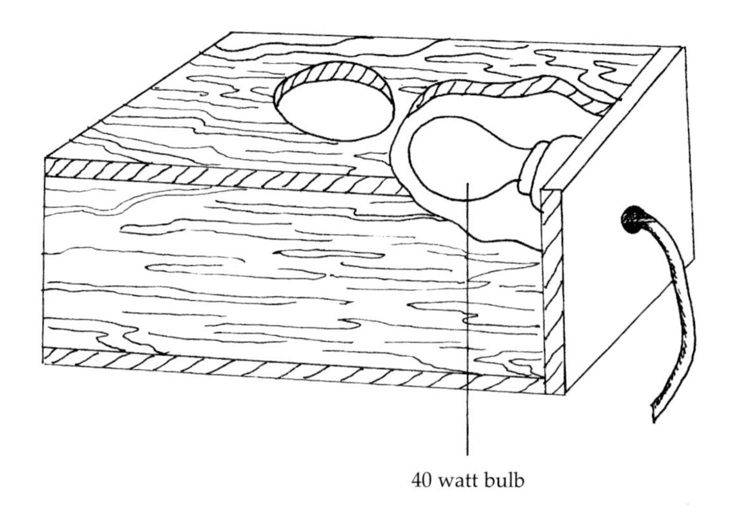

A home-made candling box.

area, in a shed, for example, or an unused garage. Let the broody come off each day to feed and drink – though for the first few days while she is sitting very tight she may have to be persuaded to come off for her feed. To do this, gently lift her up with your hands under her feet while holding her wings. This may need a bit of practice, but it is an essential part of handling if you do not want inadvertently to damage any of the eggs. If an egg is broken, then take all the eggs out, replace any dirty nesting material, and wipe off any eggs which have become sticky with a warm damp flannel before placing them back in the nest. The only feed she will need is mixed corn: 15 per cent cut maize to 85 per cent clean whole wheat. Do not be tempted to give her pellets as these will pass through her digestive system too quickly and eggs will become soiled, and this will spoil the hatch. The house or shed temperature is not as critical as that for incubators, but it should be comfortable, neither too hot nor too cold.

Candling

On the twenty-fifth day you should candle the eggs to check fertility: have a tray ready, and remove the eggs as soon as the hen comes off to feed, taking them into a dark room to candle. There are several different candlers on the market for this purpose, but if you are confident with tools, you could make up a small wooden box with a lid. Drill or cut a hole in one end to fit a light unit with a 40watt bulb. Carve out an egg-shaped hole – that is, an egg lying on its side – in the lid just large enough to place an egg on, but not so large that it drops through. Once the light is turned on you will be able to see into the egg, and should have no trouble in identifying those which the light still shines through, and those which are black with a large, clear egg cell at the broad end of the egg. Those which are clear should be discarded, and the others put back in the nest before the hen has finished feeding. If the broodies are kept too far away from the house and the source of electricity, cup the eggs one at time, balanced between the thumb and forefinger, and shine a torch up and through them.

Hatching

Humidity levels at hatching are important and, if the weather has been cold during incubation, on the twenty-fifth to twenty-sixth day you may wish to flick a very little soft water over the top of the eggs when the hen comes off to feed. During late spring when outside temperatures are hot there will be no need for any extra water, as the atmosphere holds sufficient humidity for a natural hatch to take place.

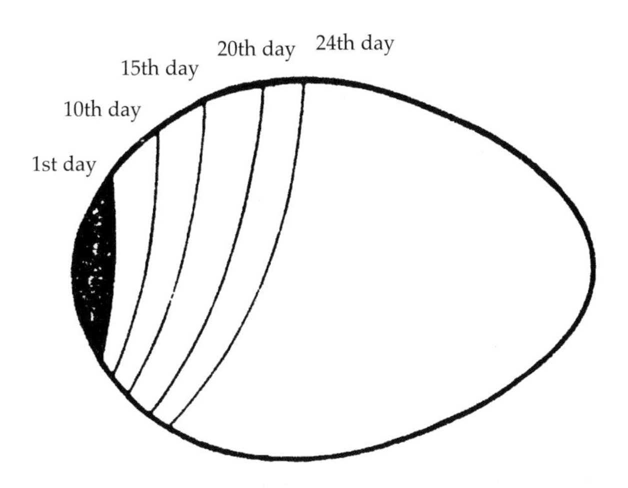

Air sac changes during incubation.

The eggs should hatch on the twenty-eighth day – that is, if you don't count the first day under the bird, as this period of time is necessary for the egg to warm up to the correct temperature in the centre of the yolk where the stag's germ is waiting to become a living embryo.

On the last day the hen may not wish to come off to feed if the young poults are already cheeping inside their shells. Do not force her off or interfere with her in any way. Leave her until the following morning, and then take a warm box to put the youngsters in as she is brought off to feed and drink. Allow her ample time while you inspect those which have hatched: any with deformities should be despatched immediately and not allowed to suffer with attempts to keep them alive. If after the full period of incubation there are still some hatching, take out the sturdy, dry poults and leave the hen for another eight hours. After this time take out any which are half hatched but stuck, or still cheeping in the shells, and destroy them, as they will be weak or deformed.

Have a clean coop ready and put the hen back in this; place all the sturdy day-old poults under her, then close the coop up for a few hours, or until they have settled down. This is more hygienic than keeping them in the same nest they have been hatched out in, which is impossible to clean out thoroughly in such a short time, thus

Broody turkeys in a corner of the house.

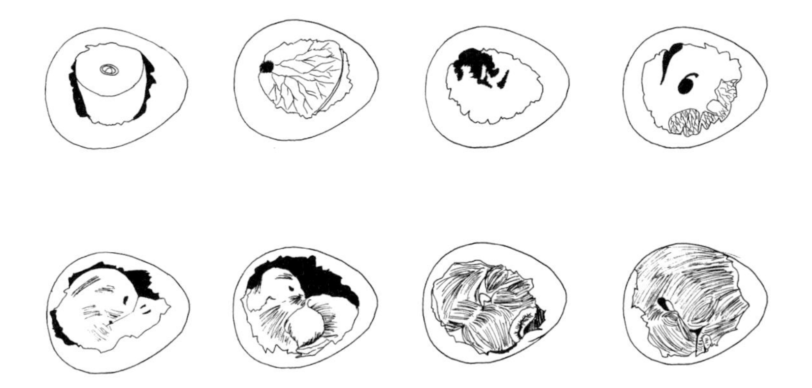

Development of a turkey chick during incubation.

increasing the risk of losses from disease. Keep all young poults away from other adult stock until they are fully mature.

ARTIFICIAL INCUBATION

Nowadays there are many different makes of incubator to choose from, and more and more firms are introducing their own ideas of what they think an incubator ought to do. It would need an extremely large chapter to explain all the innovations and novelties that are now available, however I will attempt to make the matter of artificial incubation as simple as possible, and provide the intended incubationist with the essential basics so they can then decide for themselves which system to adopt. In short, there are two basic types of artificial incubator: the 'still air' and the 'forced draught'.

The Still-Air Incubator

This type of incubator has only one layer of eggs, the air usually coming up through the base, between the eggs as it warms up, and out through small, controlled air outlets at the top. It is a very simple and efficient incubator to use; it is also usually cheaper to buy, and has very little that can go wrong, thus making it a good proposition for the small producer. Some would say that it suffers from 'dead spots', but this likelihood can be overcome when the eggs are turned – this is done manually – by taking out the eggs in the centre, rolling the eggs next to them into their place, and so on from the outside in, and finally putting the eggs from the centre on the outside perimeter of the incubator. In this way all the eggs are gradually moved about, so that all obtain as even an incubation temperature as possible. It is an enthusiast's machine, and after a few hatches the incubationist will achieve the best results possible from the fertile eggs set.

Eggs should be turned at least three times a day and, if more, the number of turning times should always be uneven. The reason for this is that if they are only turned, say, twice a day, the eggs will always spend the night on the same side. As the night is the longest period during which manually turned eggs are not moved, then the risk of the yolk becoming one-sided and sticking to the outer shell is very great, and once this happens the egg will rot. By turning three times a day, the egg will rest on a different side each night.

The Importance of Room Temperature

The temperature of the room in which the incubator is kept is of vital importance to its efficiency, and is therefore likely to affect the eggs far more than it would eggs under a broody hen. Although each incubator is thermostatically controlled inside, it cannot control the temperature of the incubator room, meaning the temperature of the air passing the incubator. This may not sound very important, as one could argue that the incubator heater is quite capable of warming the air as it enters the incubator. This is true, but what it cannot do is adjust the humidity levels at the same speed. Warm air carries more moisture than cold, so if the room temperature is generally cold, more water will be required in the incubator, and if warm, less water is required.

Room temperature is best at between 18–21°C (65–70°F); if this is maintained, then water need only be added during the last three to four days before hatching. How much is needed to suit the individual environment would be very difficult to assess with absolute accuracy, but a suggested guide would be to fill no more than one-third of the water trays or channels provided, measuring and recording the exact amount being put in. The aim is that when the poults are hatched and taken off, the water chambers should still be damp but contain no water. Any eggs not hatched must be inspected, and if the poults are fully formed but wet and sticky, then the amount of water should be reduced for the next hatch. As the season progresses and the incubator room temperature rises, then it may be that no water is required. During the whole period of incubation a daily recording must be made of the room temperature displayed on a maximum/minimum thermometer hung at incubator level in the room, and also the incubator temperature.

Room Temperature	*Incubator Temperature*
4.5°C (40°F)	40.5°C (105°F)
10°C (50°F)	40.0°C (104°F)
15.5°C (60°F)	39.5°C (103°F)
21.0°C (70°F)	38.8°C (102°F)
26.6°C (80°F)	38.3°C (101°F)

These figures are worked out on a still-air incubator, and can be used relative to those which are forced draught whose (inside) temperatures will be lower.

Forced-Draught Incubators

These incubators are different in that they can accommodate several layers of setting trays because a fan is incorporated to force the incoming flow of air through all the setting trays. Once a tray is full of eggs it restricts the natural flow of air; add another on top and air flow will be further restricted; add a third and you have curtailed

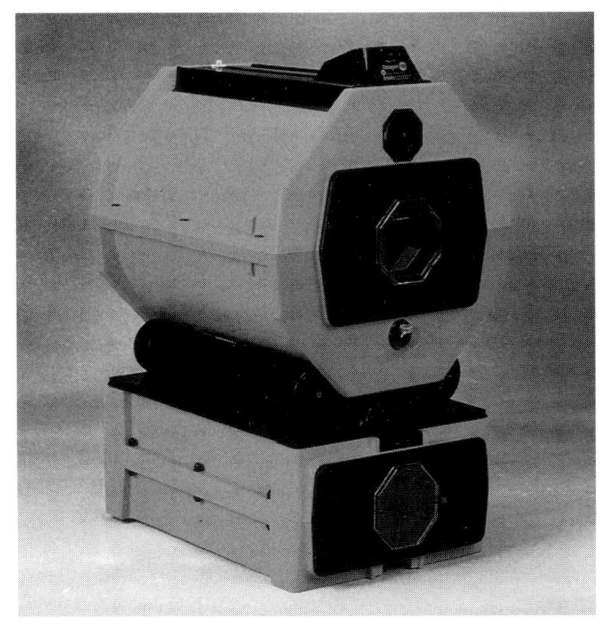

Brinsea Octagon 100 digital automatic turning and modular hatcher.

any possible chance for air to penetrate naturally throughout the incubator at an even rate. For this reason these multi-shelved incubators are fitted with a fan to force or draw the air through the eggs; this process should also prevent the occurrence of any dead spots as may be found in a still-air incubator. However, I have yet to find a forced air incubator that is capable of completely eliminating dead air spots, even the large ones of 3,000 to 4,000 egg capacity. To help overcome this problem, at weekly intervals the top tray or trays should be taken out and each tray then moved up one; finally put the top tray at the bottom. This procedure is easily carried out, even if the incubator has only been topped up on a fortnightly basis, as the gap left by the trays that are taken out to place in the hatcher will have their place filled by the eggs on the tray below, whilst the fresh eggs being set will go to the bottom.

There are incubators on the market which suggest they are capable of handling the hatching of poults whilst younger eggs not due to hatch for a week or two remain in the incubator. This is not a good idea, however, for several reasons. First, eggs ready to hatch may need to have their humidity lifted. Second, the down of the drying bird will settle on the other eggs and may partly block the shell pores. Third, as chicks hatch, so harmful bacteria are given off which may penetrate a few of the younger eggs. Fourth, the weekly temperature for an ideal hatch will vary up to 1.39°C (2.5°F) over the whole period. Finally, the hatching area at the base of the incubator needs to be thoroughly cleaned out, and this is very difficult with other eggs present at different stages of incubation. The solution is to buy a cheap, still-air incubator for hatching only. Eggs can be moved across to this four days before the hatch is due to take place, and afterwards it can be thoroughly cleaned and disinfected; then it can be left empty, giving an ideal 'disease break' before the next hatch is due to take place.

Incubator models from different manufacturers may vary in temperature because of the variations in design. For instance, in still-air incubators cool air passes in a gentle flow through the air holes in the base and on between the eggs as it is warmed up by an overhead heating unit, and its temperature will be greater than that in forced draught models where the heat is more quickly mixed with the incoming airflow as it is forced round the incubator. The variation of temperature between these two designs can be as much as 2.2°C (4°F). Most automatic units put on the market are pre-set before they leave the factory, and in most cases in a standard environment there should be no need to make any changes; however, if the poults are hatching out before the set time it will be because the temperature is too high, and a slight adjustment will need to be made. Similarly, if the hatch occurs a day or so late then the temperature will need to be increased. Never alter the temperature more than 0.27°C (0.5°F) at a time.

EGG QUALITY

There are many factors which affect the quality of incubation, but any extra cost for a quality incubator is of little use if inferior eggs are set. In the first instance, eggs for setting must be fertile, and this can only be achieved by having good, vigorous and virile stock. The stag must not be so overweight that he has trouble in mounting the hen; he must be in very good condition; he should not be inter-related, nor have a background of inbred faults; and he must be from a proven dam (hen) – that is, one that has laid well and is proven of producing healthy poults during the previous season. It is also very important that the dam produces well-shaped eggs with a sound shell.

Both the stag and the hen should be fed on a turkey breeder's ration for at least four weeks before eggs are collected for setting. If

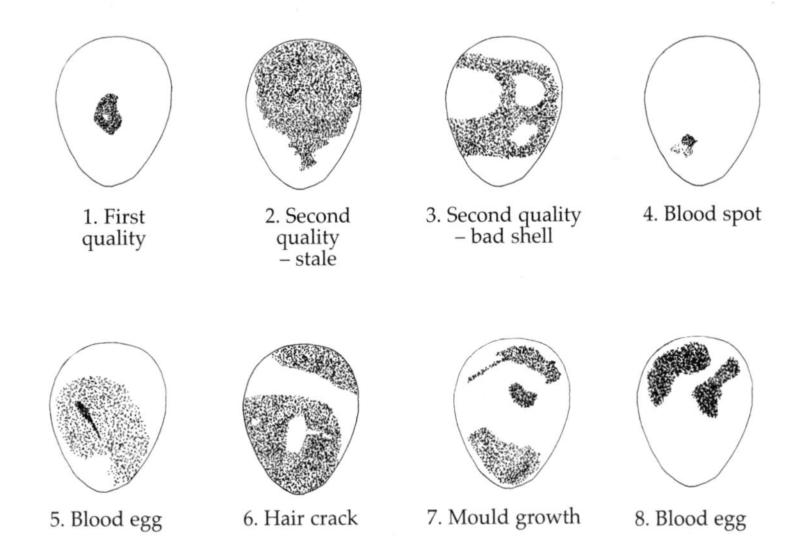

Egg quality faults.

there is difficulty in obtaining such a ration, then give soluble vitamins in the daily water at their prescribed rate four weeks before and throughout the breeding season.

Eggs for incubation should be selected for shape, shell quality, cleanliness and colour. If some of the eggs you would like to set from particular birds have been soiled, then they should be washed in water at blood temperature with added disinfectant. After washing, either run them under cold water, or place them on a draining board to cool off quickly, before putting them with other setting eggs.

A good practice is to disinfect the interior of the incubator thoroughly each time eggs are due to be put in and when eggs are moved out; this will keep harmful bacteria to a minimum. Curfew and Brinsea Incubators supply their own cabinet disinfectant: it is very easy to use and is not harmful to eggs, even if the incubator is treated just a couple of days before hatching. Remember that incubators incubate, and *whatever* is inside can multiply in the process, so the owner must ensure that it is not harmful bacteria which will prevent a successful hatch. Good hygiene is the very essence of all good incubation.

It can be seen from the table (overleaf) how vital it is to keep accurate daily records, while some of the answers on the check list are really simple common sense and can be avoided. To summarize:

1. Use only the best breeder's ration.
2. Check and record your incubator temperature once, and preferably twice, a day.
3. Check and record the max./min. room temperature each day.
4. Set only clean, fresh eggs of even size and with a sound shell, and always err on the size of caution. Don't be tempted to set substandard eggs just to fill the incubator.
5. Thoroughly sterilize your incubator between each batch.

Remember, the cause of poor hatching may not always be due to one factor alone, but to two or more. Pin your chart up by the incubator with pen attached, and fill in all the information daily. Make notes in the remarks column concerning any small details you feel could affect the success of the hatch. It is by far the best way to improve your skills and so enjoy the whole process of incubation. Don't fall into the trap of always blaming the incubator: it is rarely that which is at fault.

Reasons Why Eggs May Fail to Hatch

	Symptoms	*Usual Causes*	*Suggested Remedies*
1.	Infertile (clear egg)	Old eggs	Set within 8–14 days after they are laid
		Stags inactive	Too many or too few hens per stag
		Stags overweight	Feed only a turkey breeder's ration
		Heavy feathering round vent	Trim feathers round the vent
2.	Almost clear eggs, embryo dying early	Too high a temp.	Check and record incubator temperatures
		Chilled eggs	Protect eggs
		Insufficient or incorrect ration	Feed a quality turkey breeder's ration
3.	Poults die after eggs pipped	Humidity too high or too low	Less or more moisture, usually less.
		Temperature too low or too high for a short period	Daily temperature check
4.	Hatching too early	Average incubator temperature too high	Daily temperature check
5.	Extended hatch, i.e. some too early and some too late	Temperature too high	Daily temperature check
		Setting fresh and old eggs	Set eggs between 8–14 days. Store at the correct temperature
6.	Late hatch	Too low average temperature	Daily incubator check Possible power failure
7.	Sticky poults stuck to shell	Too much moisture in incubator	Check water left in machine. Check correct incubator ventilation
8.	Poults smeared with egg content	Combination of too low temperature and too high humidity	Daily: check incubator temperature, reduce amount of water in incubator
9.	Crooked toes	Temperature incorrect	Daily: check incubator temperature
		Genetical	Historical and visual stock selection
		Feeding incorrect diet	Feed correct ration
10.	Cross beaks	Incorrect temperature	Daily: check incubator temperature
		Feeding incorrect diet	Feed correct ration

— 6 —

The Digestive System

Before the turkey's nutritional requirements are discussed and shown in detail *(see* Chapter 8), it is important to understand its basic digestive system and the structure of its body. Most, if not all, birds have a very simple system of eating and digesting their feed to extract a small amount of protein, unlike ruminants such as sheep, cows or goats whose more efficient digestive system is able to break down and make good use even of dried poultry manure at the time of its inclusion in past rations. Because of the birds inefficient digestion, it is even more important that we fully understand the basic principles, to enable them to make the best possible use of the food provided.

Turkeys have very little sense of taste or smell, and in the main select their diet by texture and colour. If they were given a feed of wheat mixed with maize, or the same mixed in with turkey pellets, they would first selectively take out and feed on the wheat and corn because of its very agreeable texture and colour. But such a feeding system would not provide the producer with the results the birds have been bred for. They also eat oddities such as screws, nails and glass, to name but a few, so the owner must ensure that this type of debris is not left lying about.

Food is taken in between the upper and the lower mandibles (beaks) with the aid of a spear-shaped tongue, which throws it to the back of the throat; from here it is taken down the oesophagus (gullet) into the crop. It is stored in the crop until sufficiently soft to pass on into the proventriculus (glandular stomach), where the gastric juices are added before it moves on into the gizzard. This is the main powerhouse, where the grinding of all feed is carried out with the help of medium- to large-size insoluble grit, already ingested: without grit the gizzard cannot operate efficiently, as the dietary fibre, whose content is greater in birds kept on range, cannot be broken down. Even more important, all feed needs to be ground to a very fine paste, so that the maximum amount of proteins in the feed

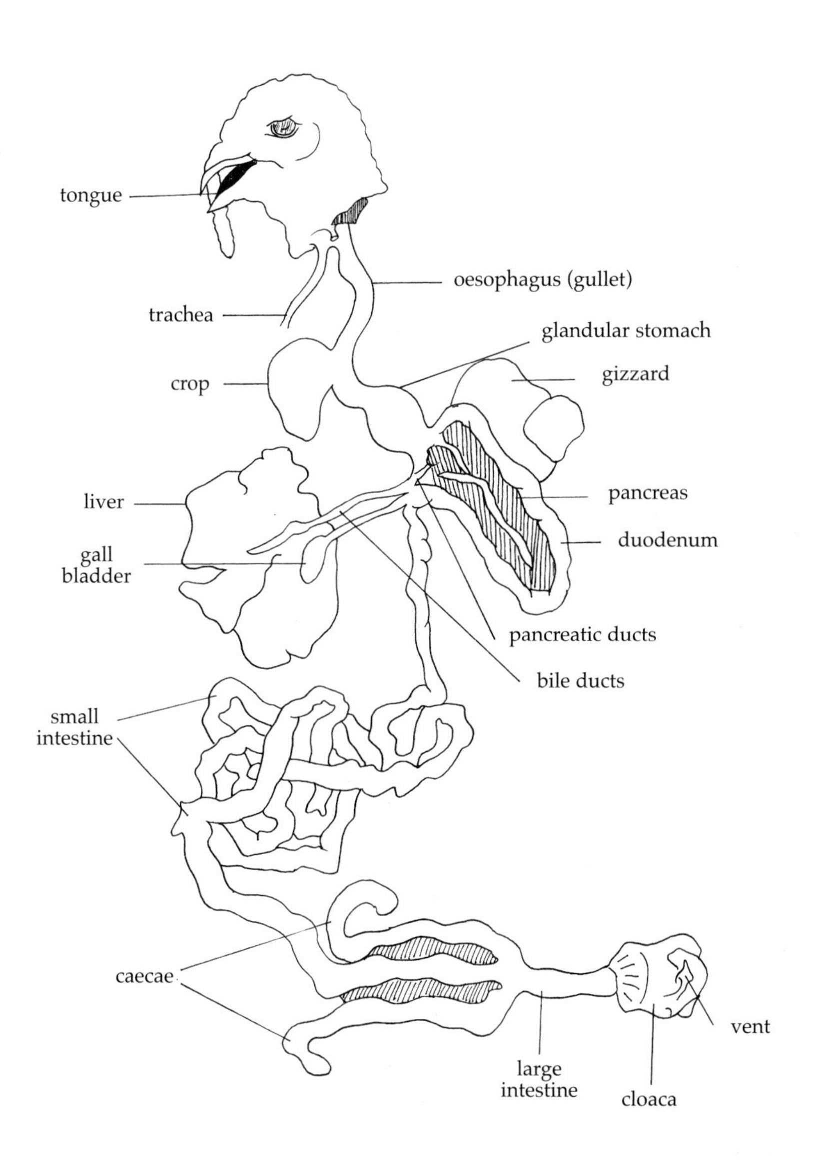

The digestive system of the turkey.

can be assimilated through the walls of the small intestine (gut). No digestion takes place in the gizzard, although a small amount of juices will pass through from the glandular stomach. The now fine-ground feed, together with water and gastric juices, passes into the duodenum where it is attacked by the pancreatic juices; these are responsible for the final breaking-down of the food constituents. The largest single organ in the body is the liver which produces bile, storing it in the gall bladder. The bile emulsifies fats which are then broken down by enzymes. The small intestine is five to six times the length of the bird and is where most of the feed is digested. The whole internal area from the duodenum to the cloaca (just inside the vent) is covered with innumerable projections called 'villi'. These enormously increase the surface area, and it is through these that the now-liquid feed is assimilated and carried by the blood to all parts of the body.

At the point where the small intestine meets with the large intestine are two blind guts called the caecae. These are more or less filled with faecal matter, and their function is concerned with the digestion of fibre and the absorption of water from the faeces.

The last part of the digestive tract is the large intestine (rectum), which is concerned with the absorption of water.

The cloaca is the part that lies immediately inside the vent, and is the area or chamber that is the common pool of the digestive tract, the ureters from the kidneys, and the genital tract; it also houses the muscle which controls the opening and closing of the external orifice.

The turkey's two kidneys lie embedded in the bony recesses on each side of the vertebrae; each kidney has three lobes that receive blood from the renal arteries and return it cleaned to the renal veins. The kidneys' function is to take from the blood all waste products: these are collected and stored in the ureters, and finally discharged into the cloaca where much of the remaining water is reabsorbed back into the body. The urine solids are excreted as the white substance capping the faeces (droppings).

The type of faeces a bird passes can be indicative of whether sufficient insoluble grit has been given, and is a significant pointer to the bird's good health. Unfortunately many producers do not pay sufficient attention to this. Daily observation can save lives, and after a time it becomes a good practical habit. As with all livestock, it is the little things that matter; the larger problems are self-evident.

— 7 —
The Structure of the Turkey

THE SKELETON

The skeleton has three main functions: to give rigidity to the body; to protect the internal organs; and to act as attachments to the muscles. These bones are composed primarily of calcium salts, and are very light to enable the bird to fly; with the exception of the bones of the tail, forearm extremities and hind limbs, they all contain air spaces.

Most of the bones in the **skull** are fused together. The **upper mandible** (beak) is movable, yet connected to the skull by the quadrate: this allows the bird freedom to open its mouth wide.

The **backbone** consists of five parts: the neck consists of thirteen or fourteen cervical vertebrae. Next to this is the thoracic region, which is relatively short, consisting of seven vertebrae. The second to the fifth vertebrae are fused together. The first and sixth vertebrae are free, and the seventh is blended with the first lumbar region. No distinction can be made between the lumbar and sacral vertebrae, about fourteen in number. The tail or caudal part consists of five or six bones, and the last few caudal vertebrae are fused into the **pygostyle** (tail); this group forms the skeleton of the tail and is capable of considerable movement.

There are seven pairs of **ribs** which protect the internal organs. The first and second, and sometimes the seventh, do not reach the sternum. Each of the other ribs consists of two segments, a vertebral and sternal.

The **sternum**, or breast bone, is long and broad, and may be described as a quadrilateral, curved plate with processes projecting from each angle and from the middle and caudal borders. It extends well towards the hind end of the bird and so helps to support the internal organs, which is necessary on account of the poor development of the abdominal muscles.

The **shoulder girdle** consists of the sword-like scapulae, stout coracoids which join the sternum, and the slender clavicles. The wing consists of a humerus, a parallel radius and large ulna, separated except at their ends, carpals, metacarpals and phalanges.

The incomplete **pelvic girdle** consists of a long ileum, flat ischium and slender pubis. The hind limb consists of the femur, tibia, fibula, metatarsal (shank) and phalanges.

The **perching mechanism** consists of a flex of tendons which curve the toes round the perch, and in the process, round the back of the tibia-metacarpus joint, therefore as the leg is bent they are tightened, causing the toes to clutch the perch so that the bird does not fall off even when asleep.

The large, heavyweight turkeys rely on receiving a balanced ration of vitamins and minerals so that they can build up a strong bone mass along with their heavier muscular body.

Identification of Bones

1. Atlas
2. Axis
3. Mandible
4. Cervical diverticulum
5. Cervico-thoracic sac
6. Anterior thoracic sac
7. Posterior thoracic sac
8. Scapula
9. Ileum
10. Diverticular of posterior abdominal sac
11. Sciatic foramen
12. Pygostyle
13. Greater abdominal sac
14. Posterior border of ischium
15. Free end of pubis
16. Lesser abdominal sac
17. Femur
18. Fibula
19. Tibia
20. Osseous tendon grove
25. Digit 4 - foot
26. Humeral and axillary diverticula
27. ... of Digit 3 - wing
28. ... of Digit 4 - wing
29. Metacarpal 4
30. Metacarpal 3
31. Ulna
32. Radius
33. Digit 2 - wing
34. Metacarpal 2
35. Ulnar carpal
36. Radial carpal
37. Quadrate
38. Jugate (xygomatic)
39. Premaxilla
40. Maxilla
41. Nasal
42. Lacrymal
43. Interobital plate
44. Coracoid

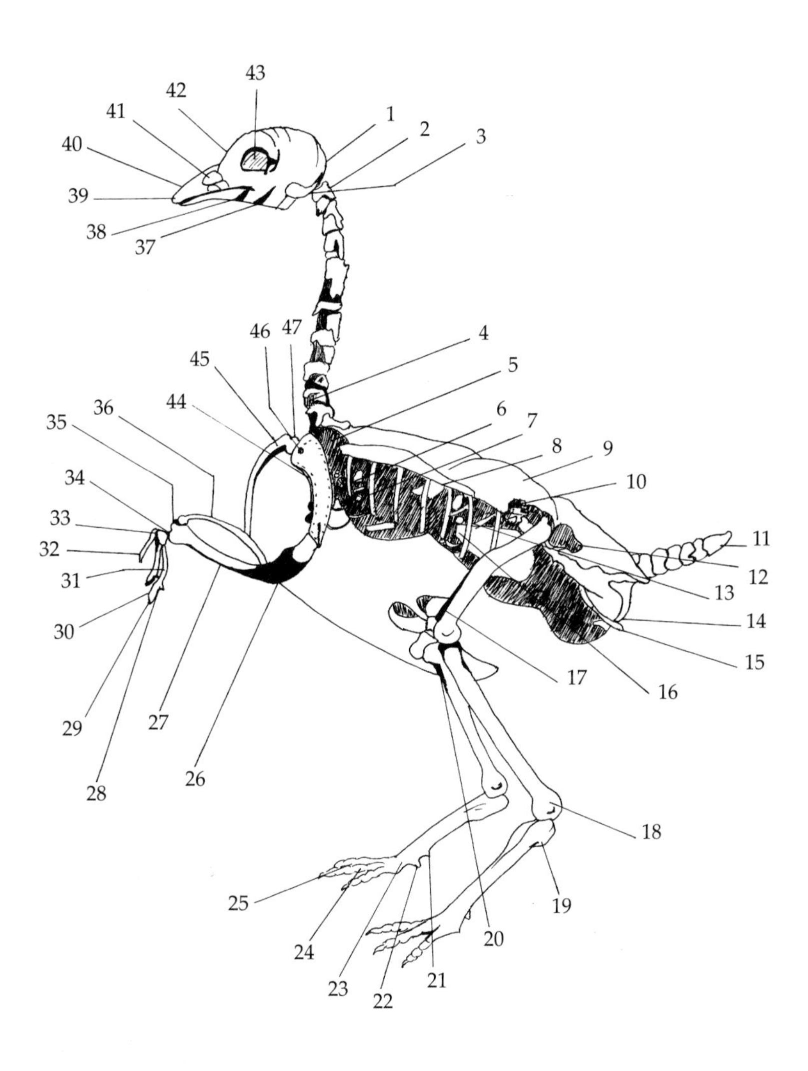

The skeleton of the turkey.

21. Bony core of spur
22. Metatarsus and digit 1 - foot
23. Digit 3 - foot
24. Digit 4 - foot
45. Clavicle
46. Foramen pneumaticum of humerus
47. Articulation of clavicle, humerus and coracoid

MUSCLES

The function of muscles is to move various parts of the body. In broad-breasted turkeys the muscles are particularly well developed and form a major part of the total weight.

SKIN

The skin includes all the facial area as well as the snood, wattle and caruncle. In the deeper layers of the skin there is muscle tissue, and this is attached to the bird's follicles; when a bird is plucked, the formation lines where the feathers were attached are easily seen. The scales on the legs and feet, and the fact that the latter are also padded, serve to protect the bird, the pads acting rather like shock absorbers.

FEATHERS

The feathers are to protect the bird, to keep it warm and dry, and to help it fly. They cover almost the entire body surface, and are of different sizes, dimensions and texture according to their function. They are all renewed annually, normally during the autumn, a process that is called a moult. Feathers which have been damaged or pulled out in the course of the year are not usually replaced until moulting takes place. Birds are at their least productive during this period, but it is important nevertheless that they are maintained on a well-balanced ration so the moult can be carried out without undue stress. Unfortunately, a few breeders, because their birds are resting over this period, only provide them with a minimum diet of wheat, or wheat mixed with corn (maize). This is very bad practice, as the turkeys require the best possible feed to produce a good coat of feathers over a normal moulting period, which takes eight to ten weeks.

— 8 —

Feeding and Weight Conversions

The nutritional requirements of any animal, let alone turkeys, hardly makes for enthralling reading, but I hope that a greater understanding of the principles of feed will help owners to become a little more discerning when purchasing turkey feed. In fact, this chapter is probably the most important in this book.

If a bird's basic nutritional requirements are not met at the more crucial stages of their development, no future standard of management, however good, will be able to compensate, and your birds will founder. Turkeys, like all other birds, require a regular supply of insoluble grit, either in the form of granite or flint grit. Without it they cannot make the best use of the food provided, whatever form it is given in, so the cost of food conversion, on a meat-to-feed ratio, will not be a healthy or viable proposition compared to those given a monthly supply of insoluble grit.

A fine grit can be introduced to three-day-old poults by sprinkling it over the starter feed, or putting a small heap at the end of each feed trough. From then on it should be made available on an ad-lib basis every four weeks: it can be put directly into the feed trough or in a separate container. Where adult turkeys or young poults are kept on a free-range basis or in large pens, it is easier to place it in a heap, changing the area each time a new amount is given. A favourite utensil for this purpose is an old tyre which can be moved around at each filling. The other advantage of providing grit at a very early age, is that it helps develop a large gizzard which will be able to deal more effectively with roughage, and in doing so the well-developed gizzard increases the weight of New York-dressed birds.

TYPES OF FEED

Starter Crumbs

These are now universally given in the UK, as no feed companies are interested in manufacturing a starter meal any more. It should go without saying that good quality rations, although on the face of it more expensive than many cheap alternatives, are usually more economical as they will produce a quality bird in a shorter period of time. That is not to say that one should buy only the most expensive: the idea is to investigate the range offered by the various manufacturers or suppliers in your area.

A starter ration will have a protein content of approximately 27–28 per cent and will include, if required, an anti-blackhead drug. When only small numbers of turkeys are kept and are not running on ground which has recently had other poultry, and where particular attention is paid to good hygiene, clean litter and alternative pens, then in these conditions there will be little need for such a drug. In some areas it is impossible to buy a turkey starter without its inclusion, and I have found that the use of an ordinary chick starter as an alternative has worked extremely well. There are those who will say that one should always stick to specially prepared turkey rations, and on a large scale I would have to agree, but many smallholders and enthusiasts find that some alternatives do work for them.

Pellets

Pellet feeding is the most common these days, not because it is necessarily the best system, but because, unless a producer is prepared to order their feed by the tonne, they will have no alternative but to feed pellets. The advantage is that there is very little wastage, poults can eat their feed quickly, and as a result the whole flock will grow evenly. The disadvantages are that by eating their day's supply very quickly, especially where birds are kept intensively or restricted in small pens, feather pecking and cannibalism may ensue.

Flocks of turkeys can generally be more aggressive than most other fowl and need to be watched very carefully. The best way round this problem is to find something else for the birds to do to relieve their boredom. The feeding of greenstuff in the form of cabbage leaves, cabbage stalks and lettuce will keep them occupied. Brassica stalks are ideal when hung up, as the poults will have great fun pecking at them and they will last several days. Largish leaves can also be suspended in a net, such as those previously containing citrus fruits – even hanging up lengths of chain will help, anything just to distract their attention away from each other.

Dry Meal (Mash)

It is now very common, for some unknown reason, for manufacturers to label meal as 'mash'. In fact the proper term for dry meal is 'meal', and the term 'mash' when used correctly refers to meal which has been mixed with water. Because of this, many of the older smallholders and farmers become confused when the name 'mash' is printed on a bag of what is obviously dry meal. This seems to have occurred at some time or other, probably when large feed manufacturers made redundant all their roving advisory staff and replaced them with *in situ* office staff. But I have included this explanation so that when we refer to 'mash', you will not expect to receive bags of wet meal.

The advantage of feeding dry meal/mash is that it takes birds far longer to consume their daily quota, so that the remaining time may be spent scratching around and resting. It must be pointed out that a minimum of twelve hours a day is needed for this type of feeding system. The disadvantage is that unless a suitable feeder is used – one specifically designed for such a purpose – then a greater amount of feed will be wasted as compared with feeding pellets in the same feeders.

If you are able to purchase a layers' meal it must be 'coarsely ground', meaning that it is quite gritty in appearance and that wheat is still visible. As has been said earlier, turkeys will eat only palatable food, and if it has been ground too fine it becomes unpalatable. However, it has been demonstrated that where birds are offered meal and pellets, they prefer meal, and are more placid and content when fed in this way.

Wet Mash

Nowadays a wet mash is only used by a small number of hobbyists who give it as an evening feed. An ad-lib supply of pellets or dry meal should always also be available. Furthermore, those who have the time and wish to give such a meal must understand that only sufficient should be prepared for a 10- to 15-minute clear-up time. Any remaining food after this period should be emptied out and disposed of. Feed contains a certain amount of yeast and will start to ferment if left, leading to probable digestive disorders. The feeders used for this purpose should be collected immediately after feeding has finished, and thoroughly cleaned. Wet mash will help weight gain, and is very much enjoyed as an extra feed by turkeys.

Its consistency is important: thus, it should not be fed with too much water, but dry enough so that when it is squeezed with one hand to form a ball, it should break open and shatter when dropped from one hand to another. During very cold days it can be mixed with warm water.

Grain Feed

This is another option, especially to small producers, hobbyists and breeders. If grain is fed it should be given during the late afternoon, and should consist of no more than 57g (2oz) per bird of a mixture of whole wheat and large cut maize. No more that 15 per cent of maize should be included in this ration.

The reason for feeding a grain mix is that not only does it cheapen the overall ration, but it encourages birds reared in intensive conditions to keep turning the litter over, which will improve or at least maintain its friability. It also encourages extensively kept birds to scratch-feed on pasture, at the same time occupying them until it has all been consumed. As birds are let out in the morning, they will often be seen returning to yesterday's feed area and working that part of the ground over again. It can help with the digestion of any heavy fibre intake, enabling the gizzard to work more efficiently.

FEED PROGRAMME

Very small numbers of poults may be started off on chick crumbs if a turkey diet is not available, but it is advisable to keep to turkey diets for larger numbers. This starter diet should be fed until they are 8 weeks of age; it is a false economy to change to a grower's ration before then. Look for a protein content of between 27–28 per cent. Although it is a seemingly expensive feed, the follow-on grower and finisher feeds are much less expensive, and it is, of course, the overall feed cost that will be taken into consideration and included in the final financial costings. All feed fed should be recorded even by the smallest producer. Records of temperature and feed will not only be vital to any poultry vet when looking at a possible problem, but will provide the owner will invaluable information as experience is gained. When only one or two batches are reared each year, it is easy to forget little items of information which could improve the proceeding year's rearing programme.

At 8 weeks of age, poults are changed to a grower's diet; this can be maintained until 16 weeks old. The protein content for this ration is 20–21 per cent. This ration can also contain an anti-blackhead inclusion if required, but must not be fed for the last two weeks prior to despatching. It is not usually necessary to give it to uncontaminated smaller flocks. The owners of a large number of small, well-maintained flocks (up to 100 poults) who do not experience problems with blackhead need only purchase a straight turkey grower's ration. Always check on the bag for any inclusions, and carry out the manufacturer's instructions to the letter. Withdrawal periods must be strictly observed.

From 16 weeks onwards either a turkey fattener may be used, or a standard adult turkey ration for birds which are to be used for breeding the following year. No inclusions of drugs should be needed in either of these rations, and check there are none.

Birds destined as breeders will not need to be given a breeder ration until about six weeks before the breeding season is due to commence. Some feed manufacturers will have five to six different rations to be fed during the growing and fattening period. These are *not*, however, all necessary to the smaller producer, only to those who produce vast numbers intensively where a single penny per pound saved in feed has some significance. When changing from one ration to another, to avoid any stress mix the two feeds together, changing to the new ration over the ensuing five to seven days. An abrupt change may cause a slight drop in consumption and so a check to the birds' system, making them vulnerable to stress, and this can lead to the dreaded vices of feather pecking and/or cannibalism. Never change the feed overnight: always do it gradually.

Feeding Table

Age of Poult	*Type of Feed*	*% Protein*
0–8 weeks	Turkey starter	27–28
8–16 weeks	Grower's ration	20–21
16 onwards	Fattening/adult ration	14–16

As water makes up the bulk of the bird's weight, it is extremely important that clean fresh water is always available.

THE COMPLETE RATION

The correct feeding of all animals is a vital part of their welfare and should fulfil four basic functions. It should:

1. Provide the right material for growth.
2. Give energy to keep the body functioning efficiently.
3. Create material to produce new tissue and to replace worn out tissue, and also to
4. Produce meat, eggs and spermatozoa.

The above functions are satisfied by feeding a complete ration, and such a ration must contain the six following substances:

1. Proteins
2. Carbohydrates (starch, sugar and fibre)
3. Fats and oils (either extract)
4. Mineral salts and minerals
5. Vitamins
6. Water

Proteins

Proteins comprise a complex group of organic compounds consisting of carbon, hydrogen, nitrogen and usually sulphur, phosphorous and iron; the presence of nitrogen is the most characteristic feature of proteins. These are body-building substances necessary for growth, body maintenance and reproduction. Proteins are said to contain approximately 16 per cent nitrogen, and in order to calculate the protein present in any substance, the analyst first estimates the nitrogen content and then multiplies this figure by 6.26. The result is rarely very accurate because the nitrogen may not all be present in the form of protein; it is therefore expressed as 'crude' protein, and it is the *crude* protein level which is printed on the bag label, not the *available* protein.

The accuracy of this figure will depend to a great extent on the constituents included in each manufacturer's formula, and it is because these materials vary between manufacturers that the formulas given can be misleading to the turkey producer; because of this, meat production or conversion figures will also vary from one brand to another. It is a general guide that very cheap rations are often made up of cheaper and inferior constituents – I say 'general guide', because better quality rations usually cost more and at the same time are more cost effective. By law, each brand of feed has to have an analysis of the feed contents affixed or printed on the bag, and some list these according to quantity, with the largest first in a decreasing order, whilst others generalize – all very confusing to the customer. There are cases where the crude protein printed on one label may actually be 2 per cent lower than another, yet the lower one may contain a higher quantity of available protein which is more beneficial to the poults. This is because the less discriminating manufacturer will include proteins which give a very high crude protein reading, of which the available protein is, however, minimal; thus he can produce a low-cost ration which will be more profitable for him to sell – but it will not give producers the results that they have every right to expect.

An old, yet sound system used by turkey and other poultry farmers to assess the value of a food, was to keep changing manufacturers over a period of time until they found a suitable feed that outperformed the others; then the producer remained loyal to that feed company's products. This method is still adopted today by some who doubt the quality of their existing ration. It is, however, detrimental to the birds' welfare if the producer keeps changing brands of feed at whim, be it for cost or convenience; this is because each firm's ration will vary in texture, and if the texture of a new feed is not as palatable to the bird as the ration it is used to, its feed intake will automatically be reduced, and this will cause stress, followed quickly by weight loss instead of weight gain.

Carbohydrates

These form a very large, diverse group of feeding stuffs consisting of carbon, hydrogen and oxygen. The hydrogen and the oxygen are always present in the same proportion as water. Carbohydrates are divided into two groups, one soluble and the other, in the form of fibre, insoluble. The soluble carbohydrates are broken down into simple sugars and absorbed to provide energy; if there is too much present in the system the excess is stored as fat. Fibre is of no use as a food, and indeed if too much fibre is present in the ration it will affect the digestibility of that ration. It has been found that rations which include more than 10 per cent fibre have a detrimental effect; however, an inclusion of fibre up to 7 per cent opens up the feed, giving it the required texture to make it palatable. Palatability is extremely important to the feeding of the bird and, should the feed be ground too fine, then the bird will not take in sufficient food to maintain itself – it may even stop feeding altogether on specific manufactured feed. The grist (coarseness) should be such that the wheat in the ration is cracked and not milled – in fact most of the main ingredients should be recognizable to the naked eye.

Fats and Oils

Whether one or the other is used is really of little importance, as both contain the same elements as carbohydrate, although they have a much greater proportion of hydrogen and oxygen. The digestive juices break fats/oils down into glycerine and fatty acids, yielding between 2.1 to 2.5 times more energy than the equivalent weight of carbohydrate.

Mineral Salts and Minerals

Mineral salts will be present on the manufacturer's label as 'ash'. To find out the amount of mineral present in a feed, a proportion of the feed is measured and placed in a suitable container, and heated until it is reduced to ash. From the residue the percentage of minerals is calculated, and when analysed it will be found to contain calcium, potassium, sodium, magnesium, manganese, iron and traces of zinc and copper, together with silicon, sulphur, chlorine and phosphorus. A bird will require many of these minerals during its life, although the majority of them are only required in minute quantities; these are referred to as 'trace elements'. They are not normally added to the diet, because they are a part of it anyway as they are present in the other ingredients.

Calcium and phosphorus form the greater part of the skeleton and egg shell. Minerals also form part of the tissue and muscles, and are necessary to maintain muscle tone and the correct pressure of various body fluids. Poultry keepers are often encouraged to feed excess oyster shell, sold separately or mixed with insoluble grit, a practice that is 'pushed' by many pet shops and cash-and-carry outlets.

With the knowledge that feed manufacturers have today, feeds are formulated to include the correct balance of calcium and phosphorus to maintain good speedy growth, and for breeders to produce eggs with good quality, strong shells. It is thought by some that calcium deficiencies can be avoided by providing ad-lib calcium. However, this practice can result in completely the opposite effect, as we shall explain: many years ago a calculation known as the 'calcium : phosphorus ratio' was established, namely that the optimum balance was 1.1 per cent calcium to 0.7 per cent phosphorus. However, should there be an excess either way, it might give rise to perosis (slipped hock tendon in poults), rickets and soft-shelled eggs. This is because both appear as tri-calcium, and if there is too much of one, it will be excreted in conjunction with the other (calcium phosphate) until only one of the properties remains, thus producing a complete imbalance.

Vitamins

It was established very many years ago that rations which included proteins, carbohydrates, fats and minerals were not sufficient in themselves to produce good growth, fertility and vitality. To achieve this, vitamins had to be added. All birds kept on genuine free-range conditions – that is, when *all* birds go out each and every day – require fewer vitamins; but where birds are kept intensively or semi-intensively, extra vitamins need to be added.

Vitamin A

This vitamin is found in fats of animal origin. A deficiency of it in a bird's diet causes leg weaknesses, general unthriftiness and white pustules in the mouth and throat; also the kidneys becomes enlarged and pale due to the collection of urates in the tubules, and it lowers the bird's resistance to disease. Vitamin A is known as an anti-infective vitamin. Birds on range cannot secure it from feeding off plants, but it is available from feed containing carotenoid pigments such as maize, alfalfa and carrots, which birds are able to convert into vitamin A. Synthetic sources are now used in most of today's rations.

Vitamin B Complex

B1, B2, B6 and B12 – which include folic acid, pantothenic acid, nicotinic acid, riboflavin, biotin and choline – are responsible for meat production. They can be included naturally in a ration via yeast, the outer coating of grains and, to some lesser extent, green vegetables. Birds kept on good pasture and given a light feed of grain in the afternoon will have an excess of this vitamin, although a winter ration for breeding birds will require extra amounts. Where this is not practised, the hatch rate is reduced, with some embryos dying during incubation. It is also responsible for curly toe in young poults, although this can be genetic or caused by uneven incubator temperatures.

Vitamin D

Calcium and phosphorus, which form the bone structure, require this vitamin so as to be able to calcify. It also assists in the hardening of egg shells. It is absorbed from direct sunlight, so free-ranging flocks will benefit. Another rich source for birds during the winter months is cod liver oil. Infra-red heaters are a good source of D3 for young and growing poults reared under electric infra-red brooders. The absence of this vitamin results in poor bone structure giving rise to brittle bones, leg weakness and poor, brittle feathering. Birds kept in a windowed house on litter will not reap the benefit of sunlight through the glass windows, as glass prevents the ultra-violet rays from penetrating.

Vitamin E

Although too much can cause sterility, insufficient amounts will lead to a poor hatch, with embryos dying during the first three to four days of incubation. Free-range stock with an extensive range will find a sufficient supply from grass and wheat.

Vitamin K

This is always included in a well formulated ration. Lack of it causes haemorrhage in the breast, legs and other parts of the body. The problem when trying to clear rats and mice from a feed store where they have access to poultry feeds is that vitamin K, being a coagulant, acts as an antidote to most mouse and rat poisons. In such a situation, try mixing drinking chocolate powder with the poison – this will ensure that they feed almost exclusively on the prepared poison and are therefore relatively easy to dispose of.

General

In many, if not all, cases of stress, a course of soluble vitamins in the water over a period of five to seven days will help a bird or flock back to health, without relying on antibiotics. Antibiotics are invaluable when prescribed and used correctly, but they should never be used as a standard treatment, as unfortunately witnessed in many small animal practices.

WATER

Water is the most important factor of any diet for all poultry, animals and humans. It must be supplied on an ad-lib basis, clean and fresh each day. Water forms the major part of the body and is essential for natural body functions; if it is restricted in any way it will cause a depletion in meat production or, where breeders are concerned, egg production; and any prolonged shortage will ultimately result in severe dehydration leading to mortality. It is well documented that man can live for a far longer period on plain water than on food, and this is also true of birds and other animals.

PALATABILITY

This has already been dealt with earlier in the chapter, but it is very important to remember that although turkeys may find a particular foodstuff extremely palatable, it may not constitute a balanced feed; thus a properly balanced diet must be provided ad lib at all times, and something which is extremely palatable, such as wheat or maize, must only be fed as a scratch ration and must on no account be mixed with meal or pellets. If this is done, the birds will tend to pick out the cereals first and they will then be too full to eat the quality ration.

WEIGHT CONVERSIONS

Whatever conversions are quoted here may vary from one breeder to another, and because of this the reader should take the following figures as a guide only. Those who achieve them, and perhaps even improve on them, will have done a very good job.

Bodyweights

All the year round, turkeys are normally killed between 12 and 18 weeks of age. These periods are extended by some to supply the larger Christmas market. To obtain these kinds of figures from as-hatched poults (unsexed), then birds should be sexed as soon as possible during the growing stage and separated at not later than 10 weeks of age.

Liveweights

Females		*Stags*	
12 weeks:	4.1kg (9lb)	12 weeks:	4.5–5kg (10–11lb)
14 weeks:	5–5.4kg (11–12lb)	14 weeks:	7.7–8.2kg (17–18lb)
18 weeks:	6.4–6.8kg (14–15lb)	18 weeks:	8.6–9.1kg (19–20lb)
		24 weeks:	12.7–13.6kg (28–30lb)

Feed Consumption

Females		*Stags*	
12 weeks:	8.2kg (18lb)	12 weeks:	10.4kg (23lb)
14 weeks:	10.9kg (24lb)	14 weeks:	18.6kg (41lb)
18 weeks:	18.1kg (40lb)	18 weeks:	23.1kg (51lb)
		24 weeks:	36.3kg (80lb)

All these calculations have been worked out on a medium/heavy turkey, and can vary not only with breed but according to temperature and possible wastage.

Feed Programme

The larger turkey growers will change from a starter ration straight to a grower ration at about 5 weeks of age. Whatever manufacturer's ration you decide on, try to keep to it.

1–8 weeks of age: turkey starter
8–12–16 weeks: turkey grower
To finish: turkey finisher (not always necessary with small flocks).

Average Turkey Weight Losses

Much will depend on the skill of the person or persons carrying out the evisceration. Below is the average type of weight loss one could expect. If the reader's figures are higher, then they will need further instruction from a skilled source. The younger the bird, the greater the weight loss.

1 per cent: starvation
8 per cent: killing and plucking
15 per cent: evisceration.
This gives a total expected weight loss of around 25 per cent.

— 9 —

Diseases and Common Ailments

To be confronted with a whole list of possible diseases can be alarming, though fortunately owners of small flocks will rarely, if ever, experience problems on such a scale. However, it is always helpful to be able to look up possible symptoms so as to get a better idea of what may be wrong with your birds, whether it is just one, or several, that are affected. I have described only the better known, basic diseases here; however, if a supplier or a poultry vet has to be consulted, it is vital to provide them with full details of the number of birds affected, their age, and their housing and feeding programme. That is why it is so important to keep a proper day-to-day record. It is over-simplistic to think that a problem can be defined on the phone from a single symptom which was only noticed just before ringing for help. Some disease problems may occur as a secondary cause, and it is worth taking note of all that has happened in the past – including handling a sick bird or birds – to ascertain where or what may have been the primary cause. Without such knowledge, although the secondary cause may be treated, it will return. A good stockman will always spend some time observing the natural behaviour of the birds in his care, so that as soon as something is amiss he will pick it up immediately and, where possible, alter management procedures in an effort to correct it. Hence the importance of keeping good records, not only of feed, temperature and mortality, but of any irregularities however insignificant they may seem at the time.

TURKEY DISEASES

Aspergillosis

This is a fungal disease, and usually affects only young birds.

However, it can affect older birds too. It is associated with the respiratory system, but there may be other sites of infection, including the visceral organs, liver, eye and brain. These latter sites are very rarely affected in the smaller producer's stock, and can only be effectively diagnosed by a good poultry veterinary surgeon.

The common signs in young poults are usually observed from 1–3 weeks of age. The affected birds will appear stunted in growth and be gasping rapidly for air. They will appear lethargic and very thirsty. Mortality can vary from 10–50 per cent.

There are several reasons for this problem. One is using mouldy litter or feed; another is the inhalation of the fungal spores that are released as damp patches in the poultry house dry out. Many a farmer or farm worker has been affected in the same way by inhaling mouldy hay or straw. Affected poults must all be culled and the litter changed, at the same time ensuring that the spores have not come from mouldy feed or mouldy corners of badly cleaned feeders. There is no feasible medical treatment for this disease.

Blackhead (*Histomoniasis*)

Although blackhead can affect many types of fowl, it is better known and more commonly found in turkeys. At one time it was a very common disease, but because much more is known about it now, and drugs more widely available, it is not experienced very often nowadays. It is a mistake to run young turkeys on ground from which hens have recently been moved, because although this disease is very rarely seen in hens they may act as hosts (carriers). I have seen a young flock of pullets suffer from blackhead, having been reared together with turkeys.

Land that has been used for chickens, even if it has laid vacant for several years, should not be used, as earthworms can also act as hosts. Seek advice from your vet if you consider that there may be a problem, and use an anti-blackhead medicant such as Dimetridazole mixed in with the feed during the early part of the growing period. You will only be able to obtain such medicated rations from an authorized feed merchant, and will have to sign each time you buy a fresh supply of feed. Note that medicated feeds have to be withdrawn from the diet a few weeks prior to killing.

The symptoms begin to show some seven to twelve days after infection has commenced, and include anorexia, lethargy and bright yellow faeces. If not treated immediately, mortality can be very high, reaching its peak about a week after the first symptoms were noticed. As soon as these symptoms are recognized via a post-mortem by a competent veterinary surgeon, a prescription for treatment will be provided.

Bordetella avium (Turkey Coryza)

This is another respiratory infection that mainly affects turkeys between 2–10 weeks of age. It can be controlled by common disinfectants and bright sunlight. In most instances it can be described as a 'secondary infection' to *E. coli* and *Mycoplasma gallisepticum*, as well as stress resulting from poor environmental conditions such as overcrowding, dirty wet litter giving off excessive amounts of ammonia, normally associated with cold temperatures, or high humidity due to poor ventilation.

The symptoms are usually heard and seen at the same time, and can also be associated with other respiratory infections. Only a post-mortem can be conclusive in this case. Infected birds start to 'snick' (light sneezing), and there is an abnormal discharge through the nose via the trachea. The eyes might be described as suffering from foamy conjunctivitis.

There is at the moment no real control available by the use of antibiotics; however, producers can avoid this problem by providing a good standard of management which involves dry, clean litter, no overcrowding and good ventilation. Once an outbreak has occurred it is important that the birds are moved into better conditions, and given soluble multivitamins in the water as an anti-stress treatment over this period.

Coccidiosis (*E. adenoeides*)

Coccidiosis is certainly one of the most serious infectious diseases in young birds. It is most common in very young poults, although it can appear in older birds which have been reared on wire floors and are then placed on litter or put on range for the finishing or growing-on period. *E. adenoedes* is the most pathogenic of the turkey coccidia, and in its worst state – when it has not been recognized by the producer and treated immediately – mortality rates will be very high.

The affected birds – and there may be only one or two to begin with – sit all huddled up, refusing to eat or drink; the faeces will be bloody, and can sometimes be seen on the feeders if the disease is not caught in the very early stages. Pick up and inspect all dead birds from about 3 weeks of age by squeezing out their faeces: if these appear fluid and bloody, and if there are one or two other poults sitting around all huddled up, contact your vet immediately. If they are placed on medication the same day, then subsequent losses will be light and the infection will rarely continue more than three to four days after treatment has commenced. Place the whole flock on multivitamins for a five- to seven-day period as soon as the coccidiosis medication treatment has finished. This will help build up a quicker resistance to any further infection which may be in the background, such as *E. coli*.

A good cleaning-out programme between batches is recommended, and if this includes wetting the floor allow a reasonably long period for it to dry out. Another form of prevention is to 'dry clean' between batches by using a scraper and brush only, as this will help all young birds who start off with parental immunity to increase that immunity at a gentle and acceptable pace. This is because a floor which has been only dry cleaned will still have the disease organisms present, but at a low level, and so the parental immunity the poults inherit will automatically build up in relation to the organisms present. The alternative is that the day-old poults are put on a hygienically prepared floor, in which case there is nothing to stimulate their immunity level, and so it begins to drop, and will continue to recede – until such time as the coccidia oocysts are brought in on someone's foot which has been soiled by faeces from more adult birds. If the challenge at this time is greater than the poults' immunity level, then an outbreak with occur. Poults hatched out and reared by a broody hen have no problem in building up their immune system naturally.

The coccidia oocysts survive well in warm and damp conditions. It is therefore very important to maintain dry litter during the brooding stages. Any water overflows or wet areas around the drinkers should be cleaned out and replaced with dry litter.

Erysipelas (*Erysipelothrix insidiosa*)

This disease is not very common in small numbers of poults, and does not usually affect birds under 13 weeks of age. However, it is a notifiable disease as it is able to cross-infect with humans. For this reason it has been included here.

First indications are sometimes when seemingly healthy birds suddenly die, and the mortality rate rapidly increases. In affected birds the snood becomes swollen, and blue-purplish areas appear on the skin; some poults will become listless with swollen joints. For some reason or other it appears to affect stags more than hens, as the organism enters through breaks and tears in the skin possibly caused by fighting. The organism is soil borne, and birds kept off the ground on wire or slats are considered to be relatively safe.

Infected birds can be injected, but this is only possible if the interval between treatment and slaughter is more than two weeks. Treatment by antibiotics will help.

As stated earlier, it is very rarely, if ever, seen in small, well-managed flocks.

Fowl Cholera (*Pastuerella multocida*)

A virulent and highly infectious disease; it is notifiable in many countries. Unfortunately, the first sign is a sudden high mortality rate in an otherwise healthy flock. An outbreak may well occur in older growers, rather than in young birds. Its carriers are rats, cats and poultry from a past infected flock. Once a site has become infected this disease may reoccur again in succeeding years. The infection is likely to be passed via water troughs as it is not airborne, as with many other diseases, nor is it passed via wild birds. It is therefore vital that water is changed each day, and the water trough washed out. Adding a small quantity of Vanodine or other water sterilizer to a header tank, or putting it in each day's water, will assist in avoiding what is a relatively rare disease these days. The advantage of medicating all drinking water in this way is that other possible problems are eliminated as well.

Poults become lethargic and go off their food, they have a mucus discharge from the nostrils, diarrhoea, possible swelling of the wattles and lameness. However, these symptoms are not uncommon to many other respiratory infections, and only a post-mortem by a qualified veterinary surgeon will confirm diagnosis.

Because of the very quick onset and high rate of mortality, treatment may possibly be a waste of money, and the owner should consider culling the remainder of the flock. It is only when this disease is less acute that treatment may appear as a worthwhile alternative. In my opinion, in diseases such as this, it is better to carry out a 100 per cent culling programme so that all houses and equipment can be thoroughly cleaned and disinfected, along with a suitable disease break (period of time), before further birds are put in the same compound.

Mycoplasma gallisepticum (Mg)

Another respiratory disease; it affects the hatch and is not always immediately recognized, in fact it can only be positively identified by sending off blood samples to the laboratory. It has possibly been present in poultry for very many years, but was not treated seriously until the advent of large-scale commercial production. I am sure that many small producers have had this organism in their flocks without even recognizing the problem, blaming the incubator, the weather or even inbreeding for a poor hatch and unthrifty poults. It can be transmitted via the egg, or be passed on by wild birds, and it has been found in the ovaries of turkeys as well as in the semen of infected stags. Once one flock becomes infected, it will spread throughout the farm, and all ages of bird will be affected. Like other respiratory diseases, it may be heard after birds have gone to roost by the occasional snick or cough. There will be a small nasal discharge, evidence of which sometimes shows up on the bird's wing, as it wipes the mucus off the nostril. Birds will also suffer from sinusitis which typically affects the eyes, which appear frothy and swollen. Unless this pathogen mixes with others there is no risk of mortality, and in fact little if any clinical signs except in very young poults.

As regards treatment, take specific advice from your poultry vet as to which would be the most effective.

Mycoplasma synovia (Ms)

The symptoms of Ms are very unlike Mg in that infected poults are retarded, have swollen joints and are lame. It is still a mild respiratory infection, although it is very unlikely that the layman would be able to diagnose it as such. Poults affected by lameness and swelling in the feet and hocks will suffer a rapid loss of condition, as they will obviously have great difficulty in getting about; they will also suffer from breast blisters. Any one of these symptoms taken on its own, before post-mortem or the growing of cultures, can very easily mislead the owner and sometimes even the vet.

When buying young or new stock it is most important that the source is reliable, and has a good reputation for bird health and quality. It is only too easy to 'buy in' this disease – and many others – via the market or showground, thus infecting any stock you may have at home.

Treatment as for *Mycoplasma gallisepticum.*

Mycoplasma meleagridis (Mm) Infection

This is primarily a Turkey disease, unlike Mg and Ms which can infect other fowl such as chicken and ducks. As a respiratory infection it retards growth, and may cause crooked necks and even abnormalities of the primary wing feathers in a few birds. It is not the problem it was many years ago, because of a steady improvement in hygiene and the controlled ventilation of turkey houses. In other words, it is very rarely seen in a well-managed environment. Young poults whose growth and feathering is slow, and which show some signs of lameness, need to be investigated.

This is a disease that it is to be hoped the reader will never see in a lifetime. Treatment is the same as for Mg, but a vet's advice will be needed and should be followed closely.

Mycoplasma iowae (Mi)

Although this disease can be found in chickens, it is another that basically affects only turkeys; it can be passed on via wild birds. As in other infections, it causes a poor hatching rate – this can be reduced by up to 10 per cent – and retarded growth, and seems to be more evident in domestic flocks than in large commercial enterprises. Some infected eggs do hatch, but the poults from these eggs exhibit poor growth and feathering. Twisted legs may be a sign of the problem – though again, this could be due to inbreeding. A great number of small breeders who experience a poor hatch as a regular occurrence accept it as a matter of course, rather than investigating the possibility – because of their own inexperience – that some of their adult stock may be carrying an infection. Those who fall prey to this complacency are generally those who do not keep any hatchery records – and as we have seen, record keeping is a must for any enlightened enthusiast attempting to improve his technique and, of course, breeding stock.

Treatment as for *Mycoplasma gallisepticum*.

Newcastle Disease (Fowl Pest)

This respiratory disease is now very rarely heard of in the UK, basically because it has been eradicated by a compulsory vaccination programme. I mention it only because one of the characteristic symptoms is for the bird to stretch its neck and gasp for breath, and this type of stretching is a symptom of other diseases, too – poults gasping for breath can be suffering from many respiratory problems and ailments, such as gape worm, or an enlarged liver, or they are simply over-fat; but the deciding factor for this disease will be the swift onset of death. Thus the few young poults that are gasping for breath and huddled up on one day will be dead the next, and the numbers that exhibit the same symptoms and die within twenty-four hours will double on each ensuing day. In my experience, the most damning symptom of this disease is this rapid mortality rate.

If the producer is at all suspicious he should ring the vet for advice, who in turn will ring the Ministry vet.

Turkey Rhinotracheitus (TRT)

The symptoms of TRT are very similar to turkey coryza. Most flocks in the UK were affected by TRT between July and December 1985, apart from those kept in Scotland. The reason for such a widespread problem was that the British turkey at that time had no natural resistance when the disease was introduced via one of the many countries in which it is standard. It comprises an acute infection of the upper respiratory tract, and mostly affects flocks between the ages of 3 to 10 weeks. Mortality can be up to, and even exceed 50 per cent, although this increased death rate is when it is associated with a secondary infection. The speed of infection is extremely rapid – akin to that of Newcastle Disease – and must be treated as soon possible after the first symptoms are seen and diagnosed.

The most noticeable symptoms are lethargy, gasping for breath, a foamy discharge from the eyes, coughing, voice change and a swollen head. The tell-tale signs are less in poults over 10 weeks of age, or in those kept on free range or in generally extensive conditions. Because of its similarity with other respiratory diseases, there is added incentive to act immediately.

It is said that only susceptible birds are infected, but then the reason for any bird going down with any infection is because it is susceptible to that particular disease. Birds kept in large units are far more at risk that those in small flocks, and it is important to note that small flocks of turkeys kept in the area of, or surrounded by, large commercial turkey units may be at risk from cross-infection.

Work is continuing to produce safe live vaccines; however, it is to be hoped that readers who keep poults in a good hygienic environment will not have the misfortune to suffer from this particular problem.

Salmonella pullorum (or BWD – Bacillary White Diarrhoea)

Pullorum was at one time a significant killer of young chicks and poults; however, as a result of the government's 'Pedigree Testing Scheme', since the early 1940s this disease has been virtually eliminated. Even so, untested stock and wild bird carriers will always pose the risk of it emerging again in small pockets in the UK.

A heavy mortality will occur in poults around 5–7 days old, although I have witnessed it in chicks from 2–3 weeks of age. They huddle up as if they have been chilled, excreting a white diarrhoea (hence the common name 'bacillary white diarrhoea'). The ovaries of infected females transmit this disease through the egg to the poult; transmission to other eggs in an incubator can also be greatly increased via the down of infected young – this alone constitutes a very important reason to disinfect between batches, and to conduct every hatch in an alternative incubator.

Once this disease has occurred, all young should be destroyed and all breeding females blood tested, because this is the only way that the breeder will be able to eliminate the carriers from the flock. Because of the rarity of this disease, there is no longer compulsory testing for its presence.

Salmonella Infection (Paratyphoid)

There are over 2,000 different strains of *Salmonella*, some of which can infect humans. There is also the alternative, yet contentious argument that humans transmit a few of the *Salmonella* strains found in poultry, such as *typhimurium* and *enteritidus*, as these are normally noted in the human population some ten years before poultry are similarly affected. It is most common in poults under 2 weeks of age and is very infectious, causing mortality of up to 100 per cent. The symptoms are common to other diseases, as affected poults stand about huddled up and with feathers ruffled, having lost all interest in eating and drinking. They may have 'pasty' vents caused by loose, sticky droppings.

As soon as the clinical signs appear, treatment can be given to reduce mortality by adding Baytril – the drug of choice – to the water over a period of ten days. There are other effective antibiotics available, and sensitivity tests will determine the best one to use. Baytril is associated with human antibiotics, so there must be a strict observance of dosage, and (even more important) of withdrawal periods. Only birds which are still eating and running about can be helped. Although treated birds may seem to recover fully from this disease, they will probably continue to be carriers, so all affected birds will need to be culled, preferably after attaining their market weight, and all housing must be thoroughly sterilized before other birds are purchased. Do not keep back any potential breeders, no matter how fit they may seem, and do not give Baytril at a reduced level as a matter of precaution, as is the unfortunate practice of some of the large commercial organizations.

Turkey Pox

Owners of well kept, well feathered, contented poults will have little to worry about concerning this disease. The virus can only infect and operate through broken skin and wounds, and will only spread from one bird to another by direct contact. It can also be transmitted via biting insects such as mosquitoes, but this is not usual in the UK. It is worth noting that when intensive production superseded extensive production and became the norm, fowl pox appeared to be a thing of the past. However, in recent years, with the return to more extensive methods, and especially because of the swiftly increasing number of birds kept by unskilled hobbyists and small producers, who provide a badly managed, unhealthy environment, this disease seems to be once more on the increase, in small and large flocks of mismanaged birds.

Symptoms are these: one or two lesions will appear on a damaged snood or wattle, followed by many more, until the whole area of the head becomes revoltingly distorted. I have even seen a case – a chicken – where the toes have dropped off, because they were in contact with the infected head area through scratching. In human terms, the best description would be a build-up of pus-filled scabs. Most of these lesions are found in the mouth, but there are others as far down as the larynx, trachea and oesophagus – and when this happens, obviously birds have great difficulty in eating or breathing. Mortality can be as high as 50 per cent, and to avoid even the possibility of this happening it is very important to separate the stags from the hens as early as possible during the rearing period to prevent infighting and immature attempts at mating.

Although there is no satisfactory treatment, there is now a live vaccine available; this is best given in the wing stub for overall vaccination. If there is no history of this disease on the farm or in the surrounding area, there is little point in vaccinating your poults. Possibly the best policy to adopt is this: when a poult is spotted with head lesions, isolate it immediately, and as soon as possible have it checked by your vet.

NOTE: When giving any type of antibiotic, use only what is recommended, and ensure that the advised withdrawal period is strictly observed.

INTERNAL PARASITICAL AILMENTS

Caecal Worms (Hairworms)

These are found in the blind gut, and an infestation will gradually build up without being noticed; the first sign of it may be that a bird dies from another cause, but one which has been triggered off by the hairworm. This worm is so small that when an examination of the faeces in the caeca needs to be carried out, the excrement has to be put in a small jar of clear water and well stirred, and it is only then that one can see these tiny worms.

As all other parasitical worms, hairworms can be treated by a seven-day dose of Flubendazole mixed in with the feed. They cause little problem in short-term flocks placed on rested land, but they may become a risk when turkeys are kept in the same run over several years, when the ground will become 'fowl sick'.

Large and Small Roundworms

Roundworms lay their eggs in the bird's intestines, and these eggs then pass out through the faeces. After a period of maturation last-

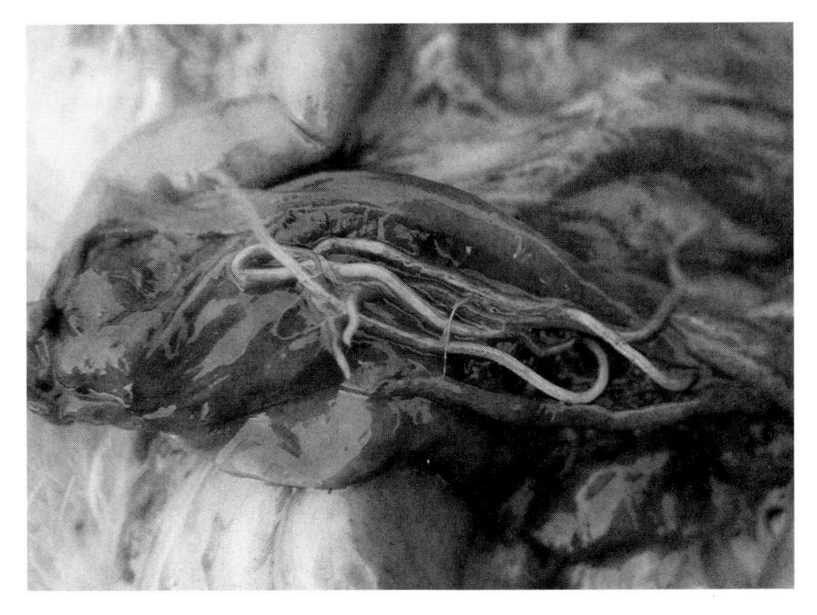

Large roundworms.

ing a week or more, they may be picked up by another bird; they will hatch in the intestine, and here develop into a mature worm. As the poults become older so the number of worms in each bird increases. The worms absorb proteins and nutrients, and as a result infested birds develop an increasingly voracious appetite. However, the head becomes paler and a poult's growth rate will come to a veritable standstill; older breeders will lose egg production. In extreme cases these worms have made their way into the oviduct and into the egg before the shell has been formed.

Treatment is as for the hairworm.

Gapeworms

The symptoms for an infestation of gapeworm may be construed as indicative of a respiratory infection, since affected poults stretch the neck skywards to try and clear the airway to be able to breathe. This problem is rarely seen in any poultry unless they are put out on infected land, possibly land which has become diseased by wild pheasants running around. The actual worm is small and pink, and its habit is to attach itself to the walls of the trachea (windpipe) thereby severely blocking the airway – hence the turkey's compulsive neck-stretching. The gapeworm's eggs pass out in the droppings, and are then eaten by a new host, the common earthworm, in

Gapeworm.

which the larvae eventually hatch. Birds are then re-infected by eating the earthworms.

To help ease the airway, and as a way of checking that the bird does in fact have gapeworm, take a long primary feather, oil it, and gently twist it down the trachea; you will need someone to keep a firm hold of the turkey while you do this. Pull it out gently, twisting all the while, and if gapeworms are the trouble, a few may be seen attached to the feather.

Treat as for all other parasitical worms.

Tapeworms

As with other parasitical worms, the eggs are passed out via the droppings or retained within small segments of the worm, which periodically break off and are excreted. These eggs, either free or within the segments, are eaten by creatures such as snails or beetles and hatch within their host, and a stage known as 'cysticercoid' develops within the latter's body. Only if the host containing the cysticercoid is eaten by another bird will the infestation be transmitted; eggs eaten in another bird's faeces will not develop into parasitical tapeworms.

It is not normal for poults reared intensively to become infected. This may only be the case when infected wild birds are able to come into a pen where the litter is wet, attracting habitation by the hosts. Again, treatment is as for all other worms.

EXTERNAL PARASITES

Lice

Lice on growing poults are rarely experienced, and normally only occur when brooded by hens themselves suffering from a lice infestation. Breeding turkeys are more likely to suffer from this particular problem. Lice act as an irritant, as they chew or bite the skin. This can in itself be the cause of an outbreak of bullying or feather pecking, leading to the inevitable cannibalism if not checked. Lice live their whole life cycle on the bird, they are golden brown in colour and can be seen more easily round the vent area, running over the skin through the feather base. They lay their eggs at the base of the feathers, emitting a gluey substance which sticks them to the feather stem and to each other. On examination these eggs may first appear as white excreta, but when looking more closely it is easier

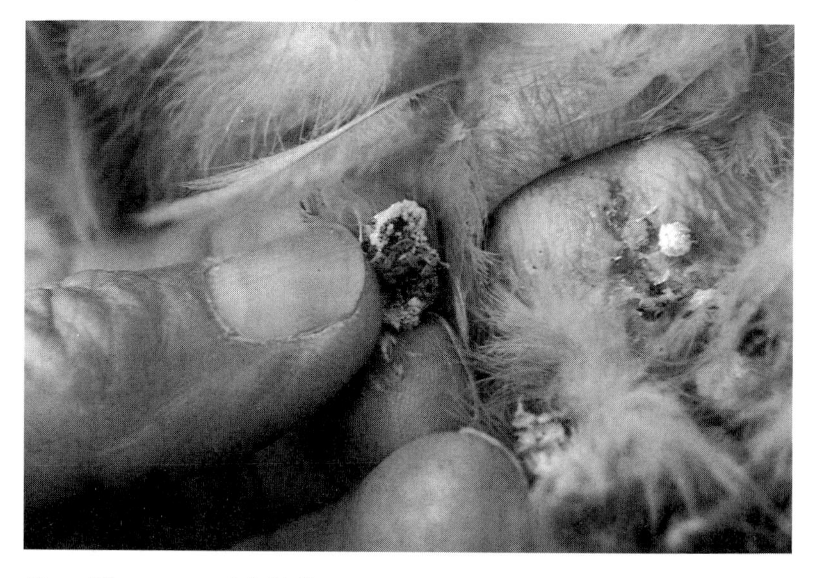

Lice and lice eggs around the bird's vent.

to see these cylindrical eggs. With a severe infestation the eggs will appear on the base of the breast feathers and under the wings. When these lice are separated from their host they live for only a short time.

Treatment is to douse the birds individually and thoroughly with a reliable louse powder. Do this again seven days later to catch new lice which have hatched. Keep nestboxes well powdered, and top dress with louse powder over the growers' dust baths. Breeding stags will have to be treated separately.

Northern Mite

These are said to inhabit the area round the vent. They are similar in size to red mite, but are grey to black in colour, and they live entirely on their host. It is my experience that these mites are more commonly observed around the head and under the wings. Birds suffering from a heavy infestation tend also to become scabby on the face and comb.

Treatment is by a spray containing pyrethrum on the affected birds. The mites are difficult to eliminate, and repeated spraying will be necessary. Lice and mites can be passed on by contact with wild birds. The greatest risk of cross infection is at shows and auctions.

Red Mite

An infestation of red mite will cause weight loss, and in breeders, a loss in egg production. According to the severity of the outbreak, if left untreated subsequent mortality will take place. The life cycle of the red mite does not depend on the host: they live and breed in crevices, as near to their next meal as possible. At one time it was thought that they lived only on wood, but they are to be found living on any material that is adjacent to the bird, including metal. Annual creosoting of the poultry house will prevent the mite from becoming established. There are now alternative sprays which will kill mite but do not remain active over such a long period as creosote.

Quite apart from causing weight loss and eventual cessation of egg production in the birds, the producer himself may experience being bitten on the head and arms when entering the house. A monthly check under perches and general house wall area will indicate any mite establishment. Only those which have sucked blood will be red; the others are grey in colour. At night they run along the underside of the perch and even over litter, and where growers are concerned, up the legs and on to the body to attach themselves to bite into a blood vessel and suck blood until they are full. They then return to the crevice and stay there until another meal is required. They do not live on the bird so will not be found by examining birds individually during daylight hours. Surrounding their crevices is a type of grey ash, which is in fact the mite faeces.

Birds suffering from the presence of a large house infestation will look very pale and jaundiced through loss of blood, and in this condition are more likely to have reduced immunity to any other infections which may be prevalent at the time. The only symptom that shows up in a post-mortem is a skin which looks as though it has been profusely pricked with a very fine needle. It is unfortunate that few laboratories will recognize the problem, and because of this the owner is left uninformed. Red mite is carried by wild birds and prevailing winds.

If old, second-hand houses are to be used, it is sensible to give them a thorough creosoting well before they are restocked. I once came across a case where an old asbestos poultry house, whose only wooden furnishings were perches, was to be restocked. There was evidence that the perches had in the past housed red mite, but the owner considered that as that had been some fourteen years ago and they had lain idle since that time, the red mite would also have died. However, within a month of birds being put in the house, there was a massive and almost immediate build-up of red mite.

Scaley Leg

This problem has all but died out amongst commercially kept birds, and is not likely to be experienced in growing poults, although it may appear on older turkeys. The leg scales become distorted as the mite burrows underneath them to nest and reproduce. As soon as this is seen, the affected bird / s must be caught and given a good leg massage with Vaseline. There are many products, most of them home-made concoctions, which claim to be effective; beware, however, as some will burn the bird's skin above the hock, and even the breast. Vaseline is extremely effective and can be applied at any time without injuring the bird at all.

OTHER MISCELLANEOUS PROBLEMS

Blue Back

This is not experienced with white turkeys, only dark-coloured ones such as Bronze or Norfolk Blacks. It is caused by continual mounting during the course of mating, not only by older breeding stags, but young stags when running with young hens which should have been separated at an earlier stage. It is caused by the stain from the continually broken back feathers being absorbed into the skin. It will not wash out, and poults affected in this way can only be sold off at a very cheap price. Turkey breeders often fix saddles on the backs of females to protect them.

Breast Blisters

Basically these are caused by poults perching on poorly constructed perches, or on hard, dirty areas of the floor; the breast blister will vary in size. It is a large area of skin filled with water and sited along the middle of the breast bone, where that part of the breast is in contact with the perch or floor, supporting to some extent the weight of the body. After plucking and just prior to evisceration, the fluid can be released by nicking the skin of the blister with a sharp knife and massaging the affected area. It can be avoided by a good system of management.

Bumble Foot

Caused by siting perches too high, or because the floor area is too hard when birds alight, especially during the late period of growth when they are heaviest. The pressure of the bird's weight on the ball of the foot as it alights swells the tissues on the underside of the central pad, so much so that the swelling continues between the toes, causing the bird to limp. It can also be caused by a staphylococcal infection if birds are kept in relatively unhygienic conditions. This latter condition is only different because the infecting swelling is hot, unlike the swelling caused by alighting from too high a perch or nestbox. Staphylococcus can be successfully treated with antibiotics supplied by a vet.

Crooked Breast

There are two main causes for this: first, it is an inherited factor – such poults should have been culled during selection at the end of the hatch; and secondly, young poults are encouraged to perch too soon, before they are ten weeks of age.

Sour Crop (Pendulous Crop)

The crop becomes over-enlarged, and very soft and squishy. The cause is a fungal infection that partially blocks the crop's exit to the glandular stomach. This will happen to otherwise healthy birds when they feed out of dirty feeders where feed has been allowed to ferment in the corners and go mouldy.

Affected birds are often successfully treated by hanging them upside down and massaging the crop while at the same time gently pushing the fluid out through the beak. As soon as this operation has been completed, poults must be isolated on a bare floor with only medicated water to drink. Vanodine is an ideal medicant to dilute, providing the correct dilution rates are observed. After the second day, feed a little corn, and the following day, if this has passed through the system successfully, gradually reintroduce the bird to its proper diet; finally, put it back with the rest of the poults.

Compacted Crop (Pendulous Crop)

This is a condition caused by birds picking up some sort of foreign material that binds up in the crop rather than passing through it – it might be the string off feed bags left lying about, long cut grass, especially when it has been allowed to dry, or perhaps feathers moulted by other birds. The crop becomes very enlarged and hard, and is generally too difficult to massage. If the condition is discovered before the crop has become totally impacted, it is sometimes possible to massage the contents out. This is best achieved by isolating the bird and keeping it in a small litter-free pen overnight with nothing but water to drink. The main problem is to remove the constriction that is preventing other feed from passing on through to the stomach. If the turkey is a prize breeder or a champion show bird, a specialist vet skilled in poultry matters may be prepared to operate to clear the obstruction; however, this will depend very much on whether the value of the bird justifies the cost of the operation. It is usually more humane to cull birds in this condition before they starve to death.

Leg and Joint Problems

If leg problems are experienced in just one bird in the flock, it is better to cull it. When there are several birds seemingly suffering from leg weaknesses, then the matter has to be explored and dealt with. In theory, no birds should suffer from rickets nowadays, and the only time this is likely to happen is when the owner is feeding too much calcium without balancing it with phosphorous, or if someone at the feed mill has pushed the wrong ingredient button. Rickets is a vitamin D3 deficiency.

If birds are fed on a manufacturer's diet, and not one concocted by the producer, then 'slipped tendon' perosis is also unlikely to be experienced. Sometimes turkeys which are free ranged are provided with only a very basic feed, being expected to balance it with all the goodies promised from a main diet of grass. However, as dairy farmers will confirm, some fields have a magnesium deficiency, and turkeys expected to survive and thrive under these conditions may also develop the problem. It can in fact be quickly remedied by additional magnesium in the feed. Those birds which are permanently crippled should be culled.

— 10 —

Organic Turkey Production

In these modern times, no book would be complete without a small chapter concerning organic production. The necessary information with regard to setting up an organic unit will depend upon which organic association you decide to join, as they do vary in how they require a producer to conform to their standards.

The Soil Association works within much stricter limits than those who adopt basic government legislation. Yet at the end of the day, producers will receive for their endeavours an identical return for their produce, even though production costs will vary between one organization and another. The basic difference between these organizations is that the Soil Association demands that all feed given to livestock contains no less than 80 per cent organic material, whereas others work within a permitted range of 70 per cent. The amount of organic material determines the price of the feed: the greater the amount, the more expensive the feedstuffs.

There is also a difference regarding house sizes and stocking densities. The stocking densities already recommended in this book are those which will provide for the best welfare of the turkey, and lower stocking densities are of no real benefit to either turkeys or producers. It could be argued that the large commercial organic free-range units would not be acceptable to someone like the Soil Association, and here I would agree. However, this book is not aimed at that type of production, but to the small producer who, I hope, is or will be very welfare minded.

It is one thing to produce organic turkey meat but another to sell it. Many poultry free-range units are set up each year under the mistaken impression that as soon as they start to sell their eggs, queues of customers will purchase all they can produce. Such producers quickly disappear off the scene, usually owing thousands of pounds to suppliers. It is no different when selling organic turkey meat, and the customers must be found and sold to before the product has even been ordered. It is also not possible to produce both normal and organic turkeys on the same premises: it is either one or the other.

REGISTRATION

To produce organic food, you will need to register with one of the organic associations operating in the UK or in the country where production is to take place. If birds are to be kept on free range, there are two things to consider: whether the soil and pasture is fit for free-range production; and the period of time one may have to wait before registration is allowed. It is normally two years, but this may be reduced if a certificate is signed to say that the land has laid fallow and that artificial fertilizers or sprays have not been used over the previous three years. This does not apply to birds being kept intensively, but it is as well to check with all such organizations concerning their requirements, as well as the cost of registration.

Registration is not free, and the cost may be too high for the number of birds you aim to produce; so check all the conditions before making a start. On first registering, you are normally allowed to sell your produce as 'Organic in Conversion', which is very helpful as an introductory period. While there are no extra costs to absorb for equipment and housing, there are for feed, which may cost as much as one third more than additive-free and other standard feeds. As feed is the major cost of turkey meat production, the cost per pound weight to cover the price of organic feed will be much more. However, committed organic customers do not appear to worry too much concerning cost – they are more interested in purchasing birds which are reared on feed produced by the purest methods of farming, and which they believe is the most environmentally friendly form of production.

IN SUMMARY

When considering taking such a step, make sure you have an organic market for all your products; then check whether you will be able to conform to, or will even be acceptable to an organic association. Find out your nearest supplier of organic feed, along with the cost, and work out how you will be able to operate in the interim period while waiting for your land to be accepted as suitable for organic production.

The market is still limited, although it grows a little each year, and this will vary from one area to another according to the earnings of the local population.

11
Despatching and Preparation

As I write there is a new type of humane killer in the throes of introduction, the brainchild of the Humane Society in conjunction with a manufacturer; they believe this new apparatus will improve killing time and so be more humane than existing methods. They are at pains to point out that instant death does not occur by simply crushing the vertebrae. For the present, however – and probably even in the distant future, too – the small turkey breeder will not be able to afford the proposed tool for killing a few birds periodically. Nevertheless, it is to be hoped that by the time this book is pub-

A pair of American white turkeys ready for Christmas.

lished, the humane despatcher will be available and bigger producers will be able to make up their minds whether it is affordable for their size unit.

DESPATCHING

At present, dislocation of the neck is still the best method of killing poultry. To do this, first stun the bird with a heavy wooden handle, and then stretch and turn the neck so that complete dislocation takes place. Continue to stretch until there is a $7^1/_2$–10cm (3–4in) gap between the dislocated vertebrae. If you find this method of dislocation too difficult, then try this way: after the bird has been stunned, lay the head on the floor and place a broom handle over the neck as near to the head as possible. Stand on the broom handle so that you have one foot on each side of the neck; then hold the bird by its legs with one hand, and the wings where they join to the back with the other, and pull up until you feel a definite dislocation, and continue pulling until the required gap is obtained. This gap is important for the blood to drain into. If you wish to bleed the turkey, then as soon as the neck has been dislocated, place the bird in a cone and sever the neck with a sharp knife. Make sure you have a bucket beneath to catch the blood.

NO ONE SHOULD ATTEMPT TO KILL TURKEYS OR ANY OTHER POULTRY WITHOUT PROFESSIONAL TRAINING.

PLUCKING

The accepted method of plucking a turkey is to tie its legs together and suspend it from a hook, thus leaving both hands free for the plucking operation. First remove the tail and the main wing feathers, then pluck the breast. After the breast has been completed, carry on to the legs and then to the back, finishing off with the wings. It is quicker and easier to clean each area as you pluck. All feathers must be removed, and fortunately there is no need to singe, as you do chickens, because turkeys do not carry body hair. Once the plucking has been completed, fold the wings back and hang the bird up in a cool room for up to fourteen days – though if the weather is warm this period of hanging will have to be reduced. Keep an eye on the neck blood: once it starts to move up the neck, then the bird needs to be eviscerated as soon as possible. The reason for hanging is to tenderize the meat and give it flavour. Once you have eaten a turkey which has been hung correctly, you will find oven-ready birds sold by the large multi-nationals soggy and tasteless.

The evisceration process is explained through the following series of photographs; it is hoped that this will provide a clear and adequate guide.

1. How to hang a turkey until it is ready for evisceration.

1. New York dressed.

2. Here the turkey is laid out on the table ready to commence work; note that the wings are now unfolded.

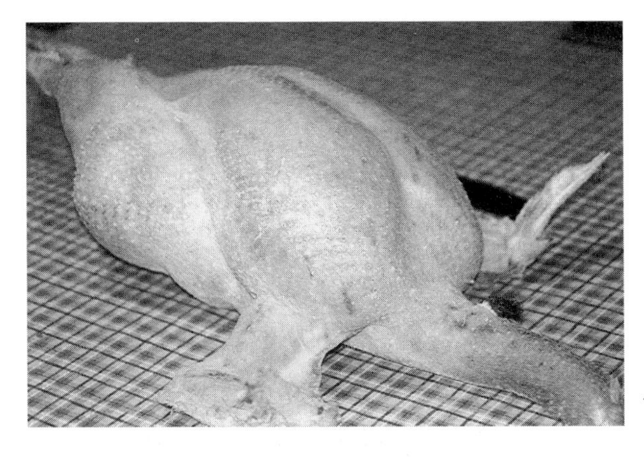

2. *Laid out ready for evisceration.*

3. First cut the skin at the hock, then twist the legs leaving the sinews ready to be extracted.

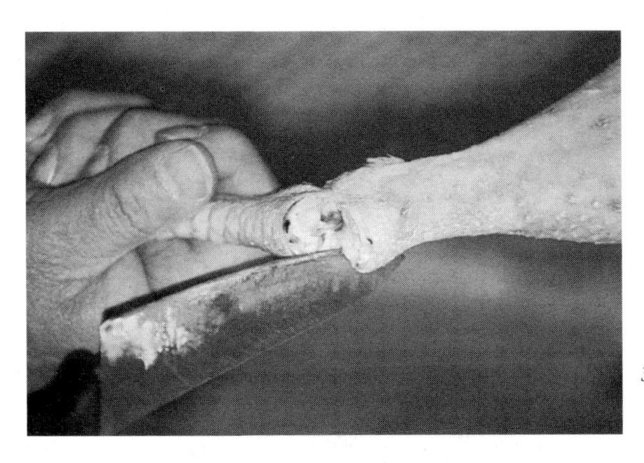

3. *Cutting the hard skin ready to draw sinews.*

4. Holding the leg firmly with one hand, with the other pull firmly in a direct line. If you find this too difficult, loop a length of binder twin or other strong string over the bird's claws, put the other end of the loop over the instep of your foot, then holding the turkey with both hands, pull up towards you (*overleaf*).

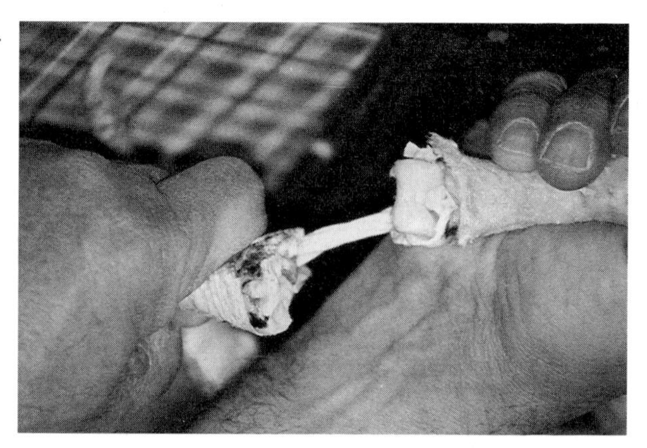

4a. Drawing sinews.

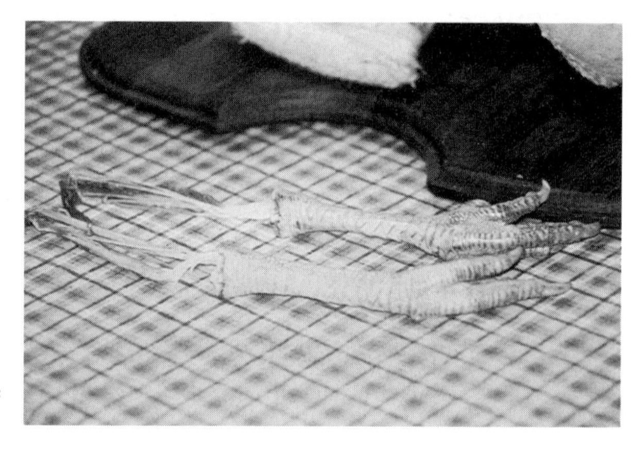

4b. Legs and sinews complete.

5. Now return to the front end of the bird, and cut off the beard (tassel) which if left is unsightly and would spoil the overall look of the finished bird.

6. Holding the neck skin with one hand so that it is tight, cut from the base of the neck to within 5cm (2in) of the head. Cut the skin just below the head, then pull the head off completely and discard. Detach the neck skin from the neck.

7. Cut the meaty section of the neck around the base, leaving it attached only by the vertebrae. One good twist will sever the neck from the body. Don't throw it away, but place it in a clean bowl.

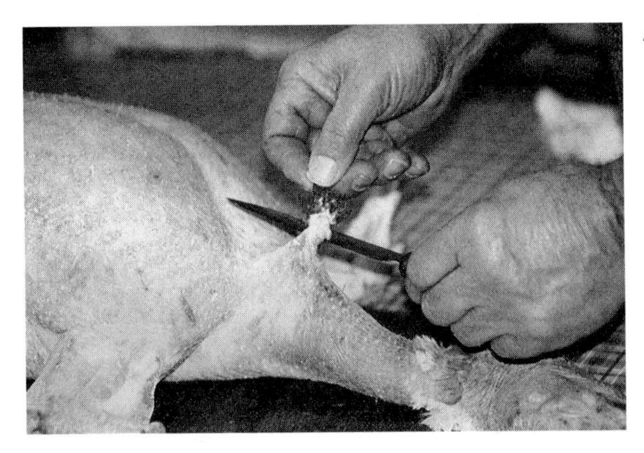

5. *Cutting off the beard (tassel).*

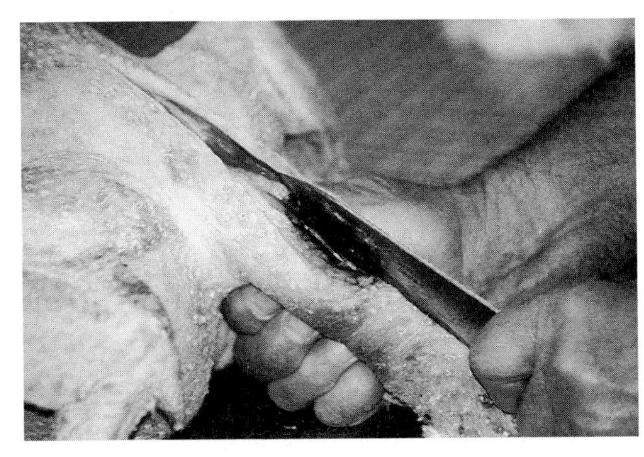

6. *Cutting the neck skin.*

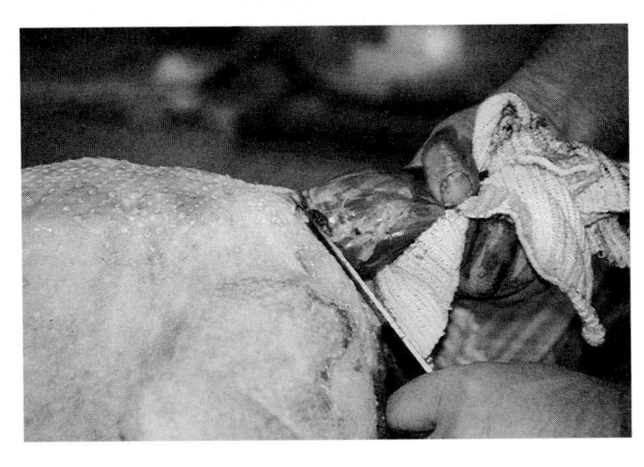

7. *Cutting out the neck.*

8. Holding the neck skin with one hand, ease away the crop and the trachea, cutting off the latter as near to the neck base as possible; discard it along with the head.

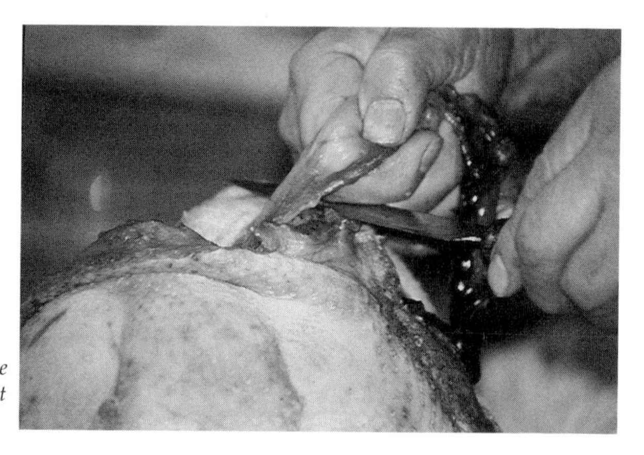

8. Cutting away the crop skin – gullet and windpipe.

9. Holding the neck skin back, ease your index and forefinger through the gap and try to separate the lungs and all other holding membranes.

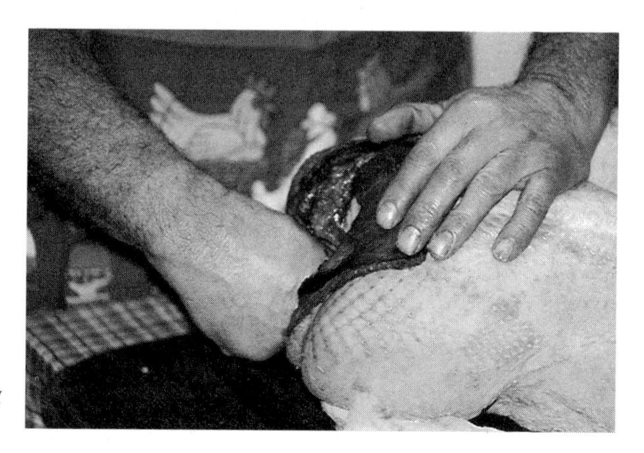

9. Loosening the lungs and forward-holding membrane.

10. Now turn the bird up the other way and cleanly cut off the oil glands above the tail.

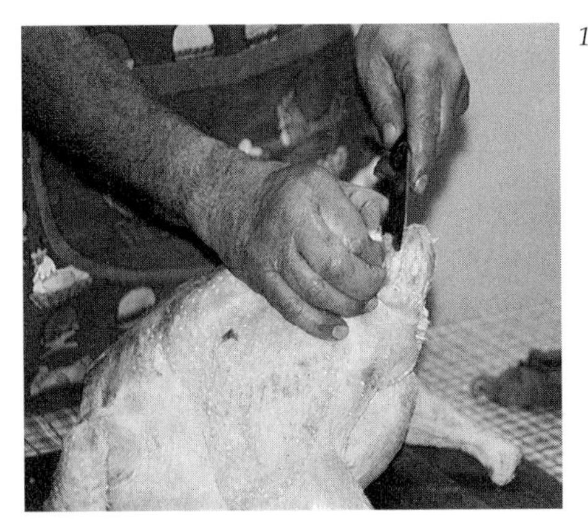

10. Cutting away the oil gland.

11. Turn the bird so that the point of the tail and back is facing you, and cut above the vent, angling the blade in towards you. Cut a hole just large enough to insert your index finger, and work your finger around the large intestine ready for the next stage.

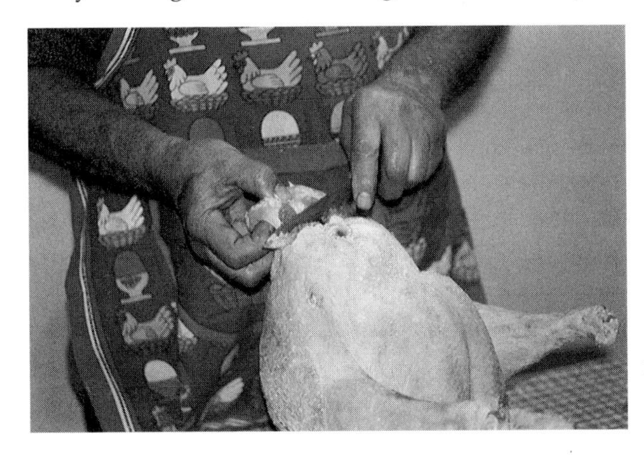

11. Cutting in towards the tail base.

12. The large intestine (gut) should still be intact. Reinsert the index finger in front of the intestine to protect it from being cut. Slide the knife between your finger and the body, and cut a large enough hole to allow your hand to ease into the body cavity; then disconnect the remaining membranes at the rear – those which could not be reached from the front end – so that all the internal organs can now be pulled out in one piece (*overleaf*).

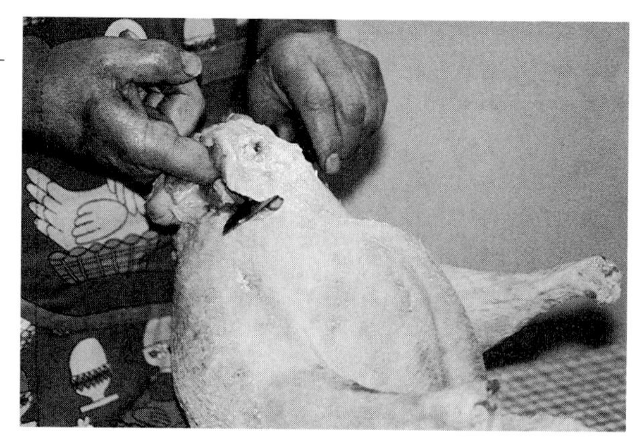

12. Finger wrapped around the gut – cutting the hole for drawing the eviscera.

13. Here you can see that the size of the hole for drawing in this particular instance is to suit a large man's hand; obviously it can be made smaller for those with smaller hands.

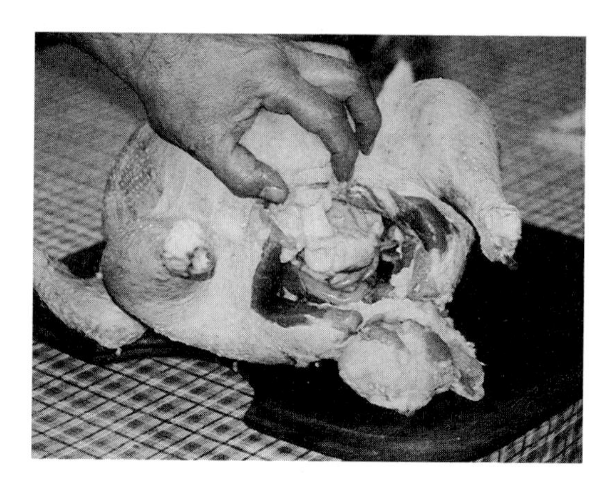

13. Exposing the opening, ready to insert hand.

14. The job finished, showing a completely empty carcass.

15. The next stage is to separate the inedible parts from the edible. Shown here is the heart being trimmed of superfluous arteries.

16. Cut the gizzard along the meaty section, just enough to expose the muscular lining.

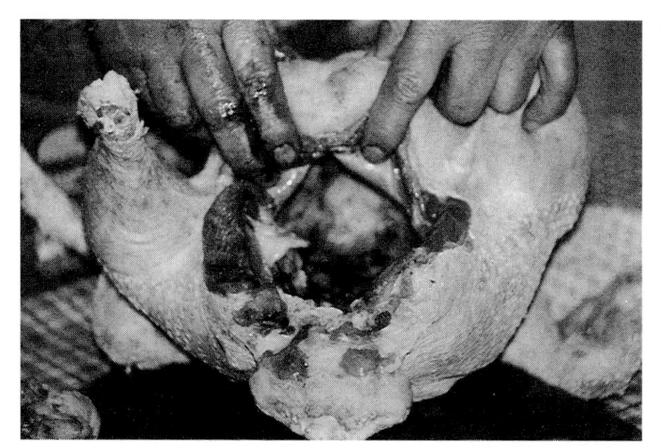

14. Cleaned-out carcass.

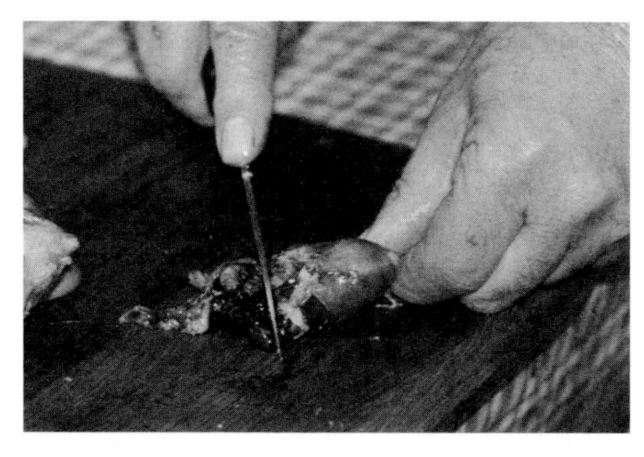

15. Cutting away the heart debris.

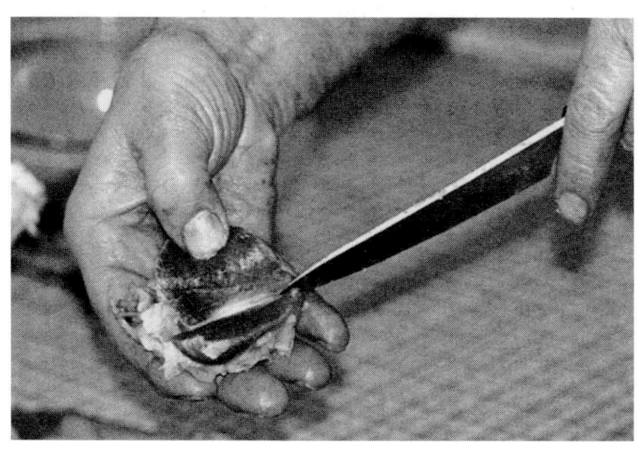

16. Carefully cutting the flesh of the gizzard.

17. This shows the cleaned gizzard, with the muscular lining separated but still intact and encapsulating the contents. The gizzard is put with the neck, heart and liver. Before the liver is added, cut out the gall bladder without puncturing it. If you do puncture the bladder by accident and the contents seep on to the liver, discard the liver because the contents of the gall bladder are very bitter. All these giblets can be packed together, but they should not be placed back inside the bird as was the traditional practice.

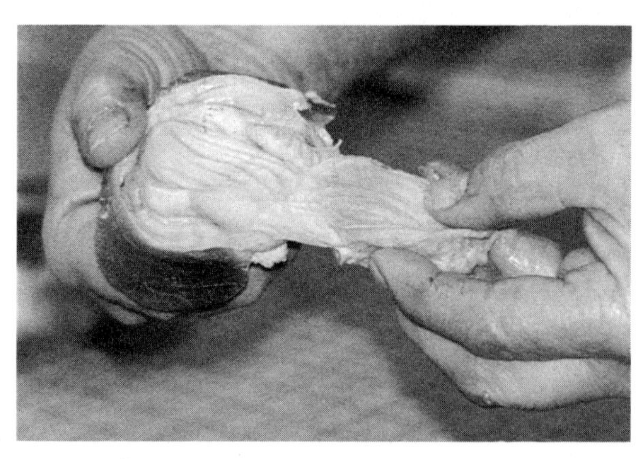

17. Separating the muscular pouch from the gizzard.

18. The carcass is ready to tie. Cut a length of new, strong string about 90cm (3ft) long; it is better to use quite thick string, as thin string will cut into your fingers when tightening. Loop it over the tail.

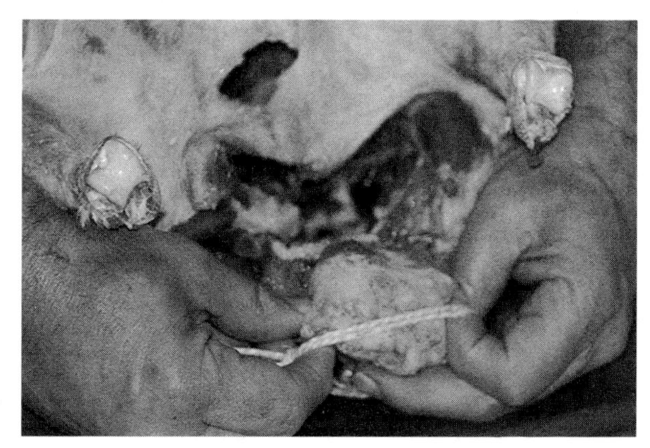

18. Tying the string around the tail.

19. Cut a small hole above the large extraction one to push the hock joints in.

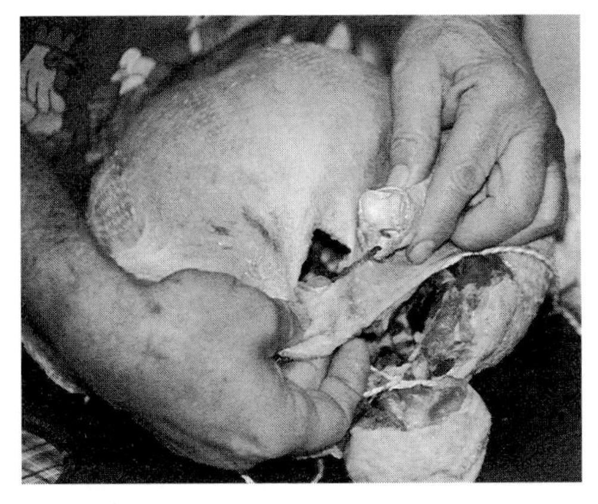

19. Putting the hocks through the small hole.

20. Tie a non-slip knot over the skin and hocks, pulling them as close to the tail as possible. This should then close off any unsightly gaps.

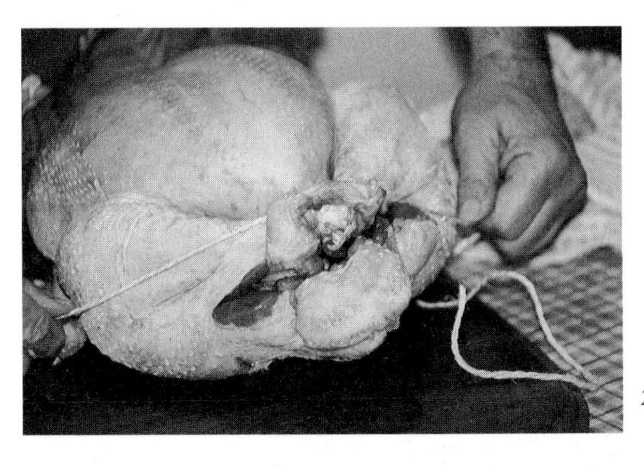

20. Tying over the skin and hocks.

21. Now pull the string tightly either side of the legs, and ...

22. ... turning the bird over on to the breast, pull the string over the front of the wings, tying tightly and securely. This in effect pushes out the breast to show off its deep meaty qualities (*overleaf*).

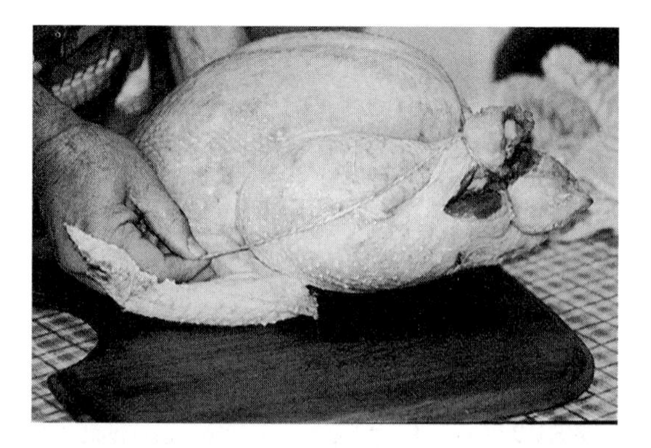

21. Drawing the string around the thighs.

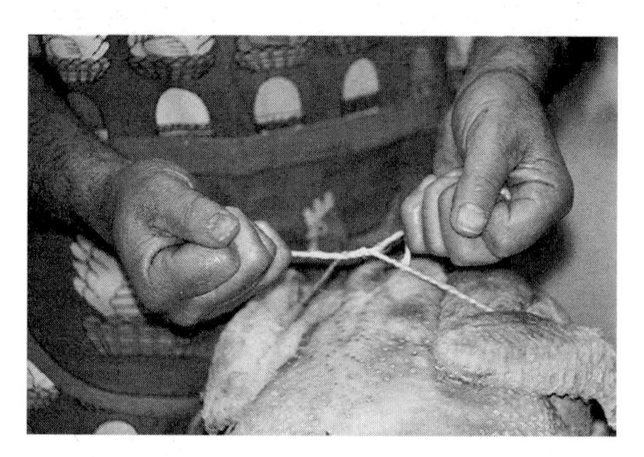

22. Final tie over the wings and back.

23. Pull the neck skin over to hide the knot ...

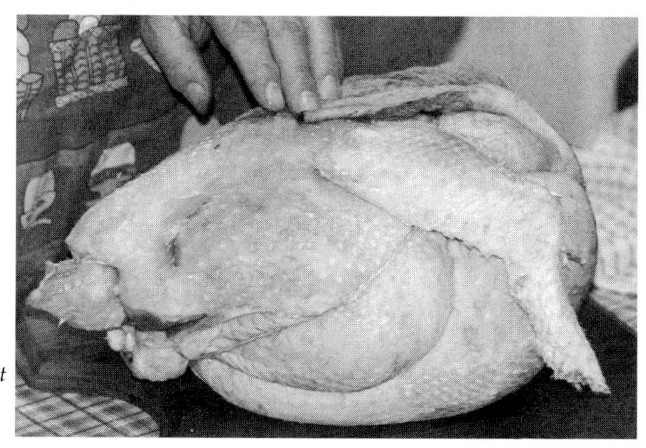

23. Covering the knot with the neck skin.

24. ... and fold the wings back to keep the skin in place.

24. Wings folded over neck skin.

25. You now have an attractively presented, deep-breasted bird ready to pack into a hygienic plastic bag. Having been hung it will not lose any flavour if placed straight into the freezer to await collection by customers at a later date.

25. The finished bird with giblets in the bowl.

If rearing and preparing a few birds for the family and relatives, provided that all these operations are carried out in hygienic conditions, there should be no problems. However, if a larger quantity of birds are reared and prepared for the table to sell to the local clientele, then the local authorities will need to be consulted first, and may wish to visit your premises to see the conditions in which these birds are to be prepared. They will then advise you if any alterations are required.

12
Codes of Practice

The welfare of turkeys can be safeguarded and their physiological and behavioural needs met under a variety of management systems. The system, and the rate of stocking at any one time, should depend on the suitability of the conditions and skills of the owner. Consideration should be given to the question of animal welfare before deciding on the system of management to be employed. Any system which is automatically controlled in any way, such as, for example, automatic drinkers, must include a failsafe alternative – in this case secondary drinkers should also be provided.

The ideal unit is one in which ventilation is natural yet baffled from the elements, and feeding and watering is manual, managed by an aware and competent owner or worker. The size of the unit should not be increased nor should a unit be set up unless it is reasonably certain that the person in charge will be able to safeguard the welfare of the individual bird.

THE TURKEY IN HEALTH AND SICKNESS

It is the responsibility of all producers to know and understand the normal behavioural characteristics of turkeys, watching closely for the first signs of distress or disease and, where necessary, taking prompt remedial action. This is more important with turkeys than with any other fowl, because turkeys will turn on a sick bird and kill it, if it has not been noticed and isn't taken out of the pen immediately by the operator, or within a very short space of time. A well-trained stockperson will know and understand all the signs which indicate good health in turkeys, and it follows that they should therefore be able to recognize impending trouble in its earliest stages – and may also be able to identify the cause and put matters right immediately. If the cause is not obvious, or if the stockperson's immediate action is not effective, veterinary or other expert advice should be sought as soon as possible.

The important indications of health are alertness, clear bright eyes, good posture, vigorous movements if unduly disturbed, active feeding and drinking, and clean healthy feathers, skin, shanks and feet. As with all other poultry, if the attendant gives a short sharp whistle on approaching, all heads should immediately shoot up to see what the noise is, and any birds which do not react like this need to be inspected more closely. I have always found this a sound way of looking for the odd sickly bird, or in fact a general pointer to the overall health of the flock.

When a whole flock is off colour, look for changes in feed and water consumption, in the general preening of the feathers, and in the amount of chattering as the birds go about their daily activities. In breeding and other laying stock, an early sign will be a drop in production and a change in egg quality, such as shell defects that are not normally experienced. Ailing birds, and any birds suffering from injury such as open wounds, fractures, or prolapse of the vent, should be separated and treated; and if they are too seriously affected or damaged for treatment to be successful, they should be humanely killed without delay.

When a bird has been taken out with a suspected prolapse, watch the other birds carefully, as what may seem to be a prolapse may well be a case of vent pecking. In this case, another bird will quickly be singled out and attacked in the same way. It is then very important to search out the perpetrator/s and separate it/them before the rest of the flock is cannibalized.

HOUSING

It is always advisable for the less experienced turkey keeper to seek advice from a prospective knowledgeable supplier as to the design and size of house that would be most suitable for the number of poults they want to rear, with special regard to their age to start with. Ventilation, heating (where necessary), lighting, feeding, watering and any other necessary equipment should be purchased to suit the number of poults envisaged and their different ages of development, and it is important that it is positioned where it will be most beneficial to the stock, and where there is no risk of injury.

All floors, and particularly slatted or wire mesh ones, should be designed, fitted and maintained so as to avoid injury or distress to the birds, and should be adapted immediately if they do. Nestboxes and perches must not be placed too high or in such a position that the birds risk injury when alighting. The best guide is that if any bird has difficulty in getting up to a perch or nestbox, then it is too high. Nestboxes should always be at least 15cm (6in) lower than perches.

The house itself should be designed for the number of birds it holds, and set out so that they all have complete freedom of movement, yet in such a way that younger or lesser birds can avoid bullies. To this end large pole barns should be equipped with rows of straw bales running nearly the width of the house, which poults at the lower end of the pecking order can escape and hide behind. The design must also be such that the operator can freely inspect his birds as they feed, drink or generally laze about, without causing a sudden panic. Even at the lowest end there must always be sufficient headroom for the largest birds, and a covered run should ideally provide sufficient headroom for the operator to be able to go about his/her duties in a comfortable manner.

All breeding birds should have a solid floor in the house and not a slatted one, as is generally the norm for growing poults.

In the larger houses or barns there should be a second, emergency door should fire or storm require that birds need to be evacuated in haste. If it is sensibly situated at the opposite end to the entrance, this second door can be used as an exit through which to pass birds as they are caught when they are to be rehoused in a range or breeding house, or for the market.

WELFARE CONCERNS

Whether electrical or gas brooders are used – and electrical ones are probably more suitable for the smaller producer – there should be an alternative standby that can be used in the event of a gas or power failure.

Nowadays, installing a fire alarm system is extremely cheap, as compared to the value of the birds. The most commonly used, and certainly the most viable for the small producer, is the smoke alarm. If these are installed, then the alarm system must be situated in such a way, or even relayed, so that the owner is alerted whether he is working on the farm, or resting at home.

The delivery of feed may not always be reliable due to weather conditions, the vagaries of the supplier, or even accident, so the manager should always carry sufficient stocks to overcome such emergencies. If owners choose to run their stock down to a day's supply or even less between each collection or delivery, then they are wholly responsible for the stress caused if they run out, and can be prosecuted should such an occasion arise.

The design of the house, whether home-built or purchased from a poultry house agent or manufacturer, should provide enough ventilation to supply sufficient fresh air for all the birds. In particular, accumulations of ammonia, hydrogen sulphide, carbon dioxide, carbon monoxide and dust should be avoided. Care should be taken to protect confined birds from draughts during cold weather.

At no time should poults be exposed to strong, direct sunlight or hot surroundings long enough to cause heat stress, as indicated by prolonged panting.

Newly hatched poults have poor control of their body temperature, so it is very important that environmental conditions during the early part of their life are sufficient for them to maintain their correct body temperature without difficulty. Whatever method of heating is used, the behaviour of poults, as described in Chapter 2, should be regarded as the best indicator of the adequacy of the environment. Young poults should not be subjected to conditions which cause either panting due to overheating, or prolonged huddling and feather ruffling because they are cold. After about four to five weeks birds can tolerate a fairly wide range of temperatures, but every effort should be made to avoid creating conditions which will lead to chilling, huddling and subsequent smothering.

Birds kept in close confinement – not that the author would recommend this – may be unable to maintain their normal body temperature; moreover, when birds are housed too closely together, the temperature may hot up enough to cause panting, particularly when the humidity is relatively high. All turkey accommodation must be designed so that even when fully stocked with turkeys at their nearly full or full growth size, the ventilation is adequate to protect the birds from all weather conditions that can be reasonably forecast.

Turkeys kept in free-range conditions must have sufficient shelter, either man-made or from natural trees and shrubs, to protect them from all variable weather conditions.

Those who wish to keep turkeys intensively – because they are short of space, or perhaps as a protective measure – should ensure that all birds receive a minimum lighting period of eight hours in every twenty-four. Enough light should be available to enable all birds to be seen clearly when they are being checked (inspected).

Turkeys, like other fowl, investigate all things with their beaks, as they have very little sense of smell or taste. For this reason all electrical installations should be positioned well out of reach of even the largest and most agile birds. The stocking rate and all other matters concerning the best environment must be taken as a whole to provide the best conditions possible, as no one item will in itself be sufficient.

STOCKING RATES

Whatever the type of closure or system of management used, all turkeys should have sufficient freedom of movement to be able, without difficulty, to stand normally, to turn round, and to stretch their wings; they should also have sufficient space to be able to perch or sit down without interference from other birds.

It cannot be too strongly emphasized that birds kept under any system can be prone to stress, injury and disease, if management and husbandry are not of a high standard. Within the present limits of scientific knowledge, it is not possible to relate stocking-rate welfare in any simple manner. Stocking rate is only one individual aspect of a complex situation involving such things as breed, strain and type of bird, colony size, temperature, ventilation, lighting and quality of housing. The observance of any particular rate cannot, by itself, ensure the welfare of the birds.

The following figures are a guide to the minimum available floor area per bird which is acceptable in most circumstances:

Rearing	*Per Bird*
Brooder houses	260sq cm per kg
Pole barns	410sq cm per kg
Enclosed pens	10sq m per kg
Breeding	*Per Bird*
Birds kept intensively, either hens for insemination or hens and stags for natural mating	515sq cm per kg
Males for artificial insemination	1sq m per bird
For show or breeding in individual pens	345sq cm per kg
Separated stags	1sq m per bird
In enclosed range areas	17sq m per bird (590 birds per hectare)

Note: If disease, especially respiratory, is observed, veterinary assistance must be obtained. Where vices are observed, if the rearer/breeder has neither the knowledge nor the experience to deal with such problems, then a qualified expert must be sought. It is to be hoped that such assistance will help prevent such incidents reoccurring.

FEED AND WATER

Birds should have easy access to adequate fresh feed and water at all times. Also, be sure that at any change of system – such as when birds are moved from one location to another – the poults find the feed and watering points. It is always advisable that when such changes take place, all birds involved are put on a five-day course of multi-vitamins in the drinking water.

Stale and contaminated feed or water should not be allowed to accumulate, and should be replaced immediately. In winter, efforts should be made to prevent water freezing; this can be done by floating glycerine over the surface of water provided in troughs, buckets or bowls.

Under no circumstances should birds be without feed or water for more than twenty-four hours.

GENERAL SUMMARY

It is to be hoped that in the course of reading these pages the reader has fully appreciated the factors necessary to the well being of the turkey. Here is a summary of the things essential to its welfare.

- Frequent inspection of the flock is vital, because the health of individual birds as well as the flock can vary from day to day. Birds must be inspected at least once a day. This can be carried out at feeding time, when the birds' reactions are more easily noted. It is not difficult, and it should become a habit to give the flock a general check at other times of the day when any other feeding or watering is carried out. This is even more important when birds are kept in intensive conditions, because if one bird feels off colour and is not spotted, within a very short space of time the healthy birds will pick on it, with possibly fatal results.
- Establish a regular work routine, so that the birds will become accustomed to their keeper's presence at different periods of the

day. In this way any likely stress will be kept to a minimum. Give them early warning of your approach, by whistling or knocking on the door before entering their house. Turkeys treated in this way will be much quieter, and birds kept for meat will provide maximum weights over a shorter period, while egg production is at its normal level.

- Protect birds at all times from predators such as the fox, rodents and other animals (including pet dogs on the loose).
- Never use mouldy litter, and keep litter dry and friable. Make a regular check for lice and mites.
- Clean out free-range houses regularly, and thoroughly clean out intensively kept birds between batches, allowing at least four weeks between cleaning and restocking as a disease break.
- An alternative run system will protect land from becoming 'fowl sick'. It is necessary to rest range which has been well stocked, on a twelve-monthly basis. If, because of the type of soil, your land tends to become very muddy in the winter/early spring, make arrangements for the flock to be housed intensively over this period.
- Turkeys kept for meat should not require a vaccination programme during their relatively short life cycle – nor, for that matter, should small, well managed, healthy flocks. It is really only the large scale, intensively kept birds which require a full vaccination programme as recommended by a poultry veterinary surgeon. In my personal opinion, small breeding flocks are better not vaccinated, because this is the best way to ensure that all breeders kept are not only healthy, but are not carriers. Vaccinations can mask any inherent weaknesses to disease etc., which will reappear in the young poults, either because of a poor hatch, or a higher mortality during the growing or laying period. Vaccinations should only be carried out by a competent and trained operator.
- Only those with the necessary knowledge and training should even consider artificial insemination as a means of reproduction.
- Hens kept with a stag or stags should be fitted with a saddle, made from a material that is hard-wearing and reliable; canvas is the popular choice. This protects the hen's back and sides against being ripped by the stag as he mounts her during mating. It is a good idea to trim the toenails of each breeding stag before placing them with the hens.
- Small flocks of well-managed turkeys should not need any debeaking; this operation only hides bad management techniques. If there is a problem, then standard management must be

checked thoroughly to correct it and prevent it from spreading. The most common reason is overstocking and boredom.

THE CULLING OF UNWANTED POULTS

Deformed day-old poults should be humanely despatched by a skilled poultry person. With small flocks this should be carried out by a quick and thorough dislocation of the neck. Chicks to be culled in this way should not be left lying about the hatchery for a seemingly indefinite period, but despatched as quickly and as efficiently as possible. All other unhatched poults still alive and entombed in the shells should be dealt with in the same manner.

THE HANDLING AND TRANSPORTATION OF BIRDS

All turkeys should be handled individually and as quietly as possible. If for any reason they panic, the whole flock will stampede, causing broken bones and/or bruising; this must be avoided at all costs. Such problems can be avoided by catching birds inside a darkened house; but should they crowd on top of one another in the catching pen, then they must be released immediately. When catching, hold both legs with one hand and prevent the wings from flapping with the other. Thus immobilized, place them in their crates head first ready for moving to their next destination.

Crated birds must *never* be left exposed to the sun.

— 13 —
Preparing Turkeys for Exhibition

The author of this book has had no experience of showing turkeys and has relied on Mrs Janice Houghton-Wallace to write the main part of this chapter for him. Her experience as a successful turkey exhibitor will help like-minded readers to understand the art of preparation, so necessary in the very competitive world of showing. Turkeys have been exhibited for well over 100 years, and their entry into the first English poultry show can be traced as far back as 1845. These days they are only likely to be seen at the major national shows, and perhaps at a few of the stronger regional ones which include classes for turkeys in their schedules. To see twenty turkeys at any of the exhibitions would be considered a very good turnout, but on the continent in Holland, Germany or Belgium, it is not unusual to see 80 to 100 turkeys on display.

HANDLING YOUR TURKEYS

If turkeys are going to be prepared for show conditions, then the earlier you start getting birds used to people and busy environments, the better. Travelling and exhibiting can be stressful for any bird, as well as the owner, so keeping it calm and confident when being handled is beneficial to the turkey's welfare as well as giving it a distinct advantage in the show pen. No one wants to see any bird looking stressed and agitated, pacing its pen, and obviously longing to get back to familiar surroundings.

Although it is most unusual for judges to handle turkeys, the people-friendly approach is most important. Begin handling the poults when they are very young, and gradually they will become docile, friendly, and used to being picked up. I make sure to stroke, hold or carry my show birds on a daily basis if possible, if only for a matter of moments. I talk to them as well. It can be about the latest cricket scores or the price of petrol, it doesn't matter, the main thing is that they are hearing my voice, and this then becomes a well known, reassuring and comforting sound.

When your birds become very tame they will not require any further pen training, although it might help a few to give them a little in the days running up to the show. All that is needed is a pen approximately 0.4sq m (4,000sq cm/4sq ft) complete with roof, made from hurdles or wire netting attached to wooden frames, placed where the turkey will be able to see and experience some activity. If given some treats, such as raisins or sweetcorn, the bird will soon settle down and become accustomed to unusual background noises and people walking around.

ENTERING A SHOW

Having decided to enter a show, apply for a schedule in good time and send your entry with the fees to the secretary well before the closing date. This can be at least a month before the event at some of the larger shows. Take your time in reading the rules and regulations, such as getting your birds on to the showground in plenty of time. Penning is when most birds should be at their correct location, and this can be quite early in the morning, so be well prepared. Allow a little extra time for parking the car. It is very frustrating if you turn up late, and find that the early arrivals have blocked the entrance to the show making it very difficult to unload anywhere near the pens.

Only enter the classes that you think are suitable, and if you enter two birds in the same class, as long as you enter early you have a better chance of them being allocated pens next to each other, so reducing a certain amount of stress. Late entries will have to take whatever pens are remaining, and these are unlikely to be together.

PRESENTATION

When the weather is inclement, turkeys benefit from being kept inside: a wet, bedraggled bird with a muddy tail and wing feathers is not a pretty sight, and will make it much harder to present them to show standards.

Throughout the summer your birds may be in different stages of moult. They will possibly be losing feathers by the handful, and you may wonder if they will ever recover. I'm pleased to say that the answer is yes, and in fact with new feathers your turkey/s will be even more handsome.

When keeping turkeys for showing purposes, do not clip their wings or mutilate their feathers in any way at all, because this will cost them valuable points – use higher netting to control those which persist in trying to fly out. If you want to roof over their outside pen, remember to have it high enough for them, and also for you so you can stand up inside, and do not have to crawl on hands and knees each time you need to go inside the pen.

Turkeys, especially stags, need to be mature enough to show themselves at their best. Although young birds may be well-marked and turned-out, they may not have quite the shape and weight to catch the judge's eye to win that all-important card.

It is very important when you first venture into exhibiting any birds that you look carefully at the quality of the winning competition and also, where possible, seek guidance from a friendly judge.

PURCHASING AN EXHIBITION TURKEY

When purchasing exhibition turkeys for the first time, go to a successful breeder, one who has a reputation for selling only the best quality stock, and is willing to advise you on the standard you need to achieve. To do this you will need to visit a few shows and make notes of winning breeders, chatting and making friends with other enthusiasts. The first poultry standards book was published in 1864 and has been continuously updated to the present day. While this book is very helpful as a guide, with pictures of show winners and details of size, feather colour, weights and conformation, the perfect bird has yet to be produced and there is no better way of learning about the standard than by visiting every show possible and discussing the different qualities of the various card winners.

PREPARING FOR THE SHOW

Having decided which of your turkeys you intend to exhibit, it is advisable to check that the bird is free from parasites. Although good management demands this anyway, for the show pen it is vitally important. Even though the judge may not handle the bird, if a stag is at the time crusted white with louse eggs at the base of its tail and vent feathers, he is sure to display them prominently to all and sundry. A continuous delousing programme should always be followed – don't leave it until a couple of weeks before the show, because eggs will already have been laid at the base of the vent feathers, and in picking them off you are bound to do some damage to the bird's appearance, however gently you do this. No matter how good your turkey is, it will lose valuable points, and in the worst scenario will not be judged if it has not been well prepared and presented.

Washing

Generally turkeys manage to keep themselves clean, so a complete wash before the show is not usually necessary, unless you have a white bird, in which case it may be. If the turkey requires a wash then complete this task a good week before the show. It will take this amount of time for the natural sheen on the feathers to return. Sometimes only part of the bird will need attention. The base of the back, tail and wing feathers will need checking. A grubby rear can be gently washed, rinsed and then dried with the lowest setting on the hair dryer.

A showing turkey in the bath for a wash and brush-up.

For a full wash, something like a large bath would be appropriate, but for a 'wash and brush up', stand the turkey in a baby bath in about 5cm (2in) of water. I use baby shampoo because it is mild, but poultry exhibitors use a wide variety of shampoos. The bird will be happier standing, and like this you will be able to control it with one hand whilst washing with the other hand. The head should need little attention, but be sure to keep the soapy water out of the eyes and ears. Do not scrub or rub the feathers, let the soapy water swill around them, or you will lose more feathers than you bargained for. Try to be gentle and quiet, not rushing too much, so that the bird remains as calm as possible. Brush the legs and feet with an old toothbrush, and when completely dry, rub a little Vaseline on both the legs and beak. Using a clean cloth, wipe a little baby oil over a stag's neck and caruncle to make him look even more impressive. If he is old enough to have a tassel (beard), then the tiniest dab of hair mousse gently rubbed down the beard between your fingers will highlight the colour and give a fresh look without it appearing greasy.

GETTING TO THE SHOW

Your bird will now be looking at his or her best, so be careful that the journey to the show does not spoil any of your preparations. For the bird's safety and to prevent it from being unduly alarmed, transport it in a large box or crate; a stag will need a larger and longer box than a hen. An adult turkey is probably longer than you think to the end of its tail – and the last thing you want is for the tail feathers to be scraped and messed up, or even worse, broken. A purpose-built crate would be lovely, though be careful it does not end up being too heavy for normal lifting and carrying. Also, take into account the size of your car or trailer, and be sure that your boxes will fit in. Providing the right travelling boxes is a job that can be done well in advance of the show – for your own peace of mind, don't leave it until almost the day of the show, because something is almost certain to go wrong. The large, strong cardboard boxes originally meant for transporting televisions or suchlike are ideal. Reinforce these with string or strong commercial sticking tape so that the bottom is well supported and won't collapse at an inopportune moment.

I will generally carry a turkey to a box which I have placed near the car, put it in, and then lift the box into the travelling position in the car. Once at the show I place the box on the ground, then lift the

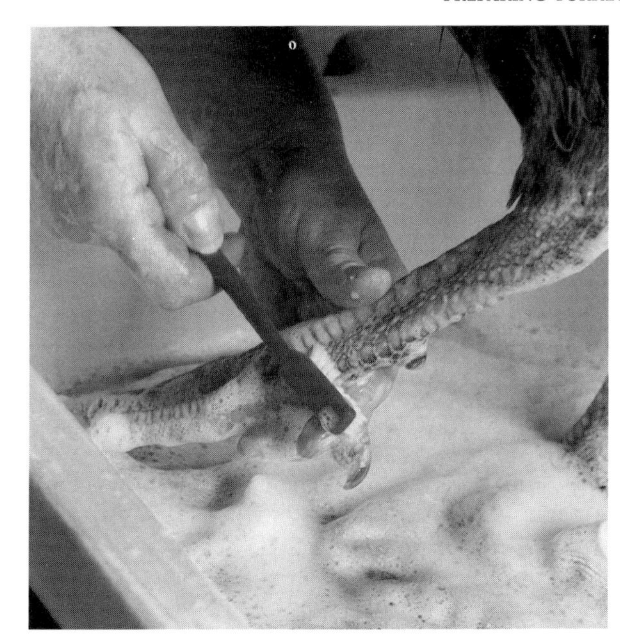

Washing and scrubbing the legs and scales.

Partial wash around the vent area.

turkey out and carry it to its pen. By all means, if you have someone strong to help with the carrying, it may be preferable to take it still in its box directly to the show pen. However, I feel I have more control over the turkey doing it the other way – it won't fall about in the box, and I know I have a firm grip of it all the time. Don't ever underestimate the weight of a turkey: they can be very heavy, and they are certainly very strong, so handling them needs practice and technique. Hold it under your arm so that it is facing behind you – like this it will be difficult for its wings to get loose and flap, when it might give you a nasty clout; use your left hand (or whichever is the strongest) to grip its legs. Any bird being carried needs to feel secure and safe, so handle it firmly but gently. However, it will not appreciate being grasped too tightly and squeezed.

AT THE SHOW

Although some shows provide water containers in the show pens, always take your own along, just in case. A large margarine or ice-cream tub with wire threaded through one side to attach to the pen frame, will provide sufficient water for your bird, and you can always check it periodically and top it up. A new container is bound to be more hygienic, and protecting your birds in as many ways as you can from disease is always worthwhile. Also, take along sufficient food for the duration of the show. Sometimes food is provided, but if a turkey is familiar with a particular ration it may cause it unnecessary stress to change to another diet – it is already experiencing a certain amount of stress in acclimatizing to its new environment. A piece of apple or sweetcorn after judging will help to keep it occupied and happy.

When you arrive at the show, collect your pen number and then find the pen allocated to you; this way you will not be exploring what could be quite a large marquee or hall, trying to find your pen with a large, heavy turkey tucked under your arm. Double check that you have put your exhibit in the correct pen, and then do the final bit of presentation by gently stroking its feathers with a silken cloth. Rearrange any feathers that need it – and then tell your turkey to do its best.

AFTER THE SHOW

After the show get your bird home as quickly as possible and give it a good delouse, just in case some parasites were picked up from other birds in the show. If it is still daylight when you arrive home, then the turkey may relish a good run around and wing-stretch. However gentle you have been with it during the show day, it will still have been quite a stressful experience, so spoil your birds for a few weeks to help get them back to full condition. This can be greatly helped by putting a soluble multivitamin in the water for the next five days.

SUMMARY

In this chapter I have tried to give as much advice as possible; however, please don't be put off by the seemingly endless tasks that are required in order to exhibit a turkey. The judging is, after all, only a guide as to the quality of your bird/s. One of the most rewarding parts of showing turkeys is the pleasure and friendships that you gain from meeting other exhibitors. There are not the number of turkey enthusiasts as there are other types of poultry fancier, so another like-minded person is always made very welcome. The

Blow-drying the vent area.

shows take you to different parts of the country, and in the process you meet knowledgeable people and learn from them as you go along. Also, it's a good idea to take a camera with you. The poultry magazines are always pleased to receive something different for their pages, and turkeys, especially those shown, do not normally get very much coverage. Should you be fortunate enough to win a prize, then congratulations, this could be the first of many.

Enjoy showing your turkeys, and encourage others to do the same.

The end result.

Index

INDEX